Network Analysis and Tourism

D1425347

ASPECTS OF TOURISM

Series Editors: Professor Chris Cooper, *University of Nottingham, UK*
Dr C. Michael Hall, *University of Canterbury, Christchurch, New Zealand*
Dr Dallen Timothy, *Brigham Young University, Utah, USA*

Aspects of Tourism is an innovative, multifaceted series which will comprise authoritative reference handbooks on global tourism regions, research volumes, texts and monographs. It is designed to provide readers with the latest thinking on tourism world-wide and in so doing will push back the frontiers of tourism knowledge. The series will also introduce a new generation of international tourism authors, writing on leading edge topics. The volumes will be readable and user-friendly, providing accessible sources for further research. The list will be underpinned by an annual authoritative tourism research volume. Books in the series will be commissioned that probe the relationship between tourism and cognate subject areas such as strategy, development, retailing, sport and environmental studies. The publisher and series editors welcome proposals from writers with projects on these topics.

Other Books in the Series
Tourism Development: Issues for a Vulnerable Industry
Julio Aramberri and Richard Butler (eds)
Nature-based Tourism in Peripheral Areas: Development or Disaster?
C. Michael Hall and Stephen Boyd (eds)
Tourism, Recreation and Climate Change
C. Michael Hall and James Higham (eds)
Shopping Tourism, Retailing and Leisure
Dallen J. Timothy
Wildlife Tourism
David Newsome, Ross Dowling and Susan Moore
Film-Induced Tourism
Sue Beeton
Rural Tourism and Sustainable Business
Derek Hall, Irene Kirkpatrick and Morag Mitchell (eds)
The Tourism Area Life Cycle, Vol. 1: Applications and Modifications
Richard W. Butler (ed.)
The Tourism Area Life Cycle, Vol. 2: Conceptual and Theoretical Issues
Richard W. Butler (ed.)
Tourist Behaviour: Themes and Conceptual Schemes
Philip L. Pearce
Tourism Ethics
David A. Fennell
North America: A Tourism Handbook
David A. Fennell (ed.)
Lake Tourism: An Integrated Approach to Lacustrine Tourism Systems
C. Michael Hall and Tuija Härkönen (eds)
Codes of Ethics in Tourism: Practice, Theory, Synthesis
David A. Fennell and David C. Malloy
Managing Coastal Tourism Resorts: A Global Perspective
Sheela Agarwal and Gareth Shaw (eds)

For more details of these or any other of our publications, please contact:
Channel View Publications, Frankfurt Lodge, Clevedon Hall,
Victoria Road, Clevedon, BS21 7HH, England
http://www.channelviewpublications.com

ASPECTS OF TOURISM 35
Series Editors: Chris Cooper (*University of Nottingham, UK*)
C. Michael Hall (*University of Canterbury, New Zealand*)
and Dallen Timothy (*Brigham Young University, Utah, USA*)

Network Analysis and Tourism
From Theory to Practice

Noel Scott, Rodolfo Baggio and Chris Cooper

CHANNEL VIEW PUBLICATIONS
Clevedon • Buffalo • Toronto

Library of Congress Cataloging in Publication Data
Scott, Noel
Network Analysis and Tourism: From Theory to Practice/Noel Scott, Rodolfo Baggio and Chris Cooper.
Includes bibliographical references.
1. Tourism. 2. System analysis. I. Scott, Noel. II. Baggio, Rodolfo. III. Cooper, Chris. IV. Title.
G155.A1N426 2008
910.068'4–dc22   2007050183

British Library Cataloguing in Publication Data
A catalogue entry for this book is available from the British Library.

ISBN-13: 978-1-84541-088-9 (hbk)
ISBN-13: 978-1-84541-087-2 (pbk)

Channel View Publications
UK: Frankfurt Lodge, Clevedon Hall, Victoria Road, Clevedon BS21 7HH.
USA: 2250 Military Road, Tonawanda, NY 14150, USA.
Canada: 5201 Dufferin Street, North York, Ontario, Canada M3H 5T8.

The policy of Multilingual Matters/Channel View Publications is to use papers that are natural, renewable and recyclable products, made from wood grown in sustainable forests. In the manufacturing process of our books, and to further support our policy, preference is given to printers that have FSC and PEFC Chain of Custody certification. The FSC and/or PEFC logos will appear on those books where full certification has been granted to the printer concerned.

Typeset by Bookcraft Ltd.
Printed and bound in Great Britain by the Cromwell Press Ltd.

Contents

Figures

Contributors*

Noel Scott has extensive experience as a senior tourism manager and researcher and over 25 years in industry research positions. He holds a doctorate in tourism management and Masters degrees in marketing and business administration and is a senior lecturer at The University of Queensland, Brisbane, Australia.

Chris Cooper has degrees in geography from University College London. He is joint editor of Channel View's Aspect of Tourism Series and co-editor of Current Issues in Tourism. His research interests lie in the area of destination management, particularly focusing on network analysis and innovation. He is currently director of the Christel DeHaan Tourism and Travel Research Institute at the University of Nottingham, UK.

Rodolfo Baggio has a background in physics and actively researches and lectures in the field of information technology and tourism. He is now applying quantitative network analysis methods to the study of tourism destinations.

* * * * * *

Zélia Breda is a lecturer and PhD student in tourism at the University of Aveiro. She holds an MA in Chinese Studies and a degree in tourism management and planning from the University of Aveiro. Her research interests include tourism development and planning, clusters and networks, tourism in Asia, international investment, tourism politics and economics. She is author and co-author of national and international publications, and has participated in various international conferences.

*Authors of individual chapters are indicated at the head of the chapters concerned. All other material has been written by the three principal authors.

Carlos Costa has been Director of the Portuguese teaching programmes of the Aveiro University since 1996 and is a visiting senior fellow at the University of Surrey. He holds a PhD and MSc in tourism management from the University of Surrey. He is also Editor of the Tourism & Development Journal, the first tourism journal in Portugal. Carlos has also been involved in several projects and consultancy work for the Portuguese Government and other Portuguese organisations.

Rui Costa is a tourism PhD research student at the University of Aveiro and research fellow of the Foundation for Science and Technology. He holds an MSc in innovation and development policies and a degree in tourism management and planning from the University of Aveiro. His research interests include tourism planning and development, tourism networks, innovation and entrepreneurship, project evaluation and monitoring.

Dianne Dredge's main areas of interest are in policy networks, and their structure and role in tourism planning and development. She also has active research interests in social constructions of tourism places, networks and development conflicts, tourism governance and inter-governmental relations. She is an active member of the planning profession and is involved in a range of tourism planning consultancies. She is an Associate Professor at the Centre for Tourism, Sport and Service Innovation, Griffith University, Australia.

Roger March's research interests include international tourism, marketing in Asia and distribution channels and systems in relation to tourism service provision. He does consulting work on Japanese and Asian tourism markets for Australia's leading tourism organisations and works as cross-cultural trainer of Westerners working in Japan or in Japanese companies overseas. Roger is a Senior Lecturer at the School of Marketing, University of New South Wales.

Giuseppe Marzano holds a doctorate in tourism from The University of Queensland where he lectures in tourism management. His research currently focuses on stakeholder interactions in decision-making processes in the context of tourism with a particular emphasis on collaboration and power issues.

Joana Miguéns is a tourism PhD research student at the University of Aveiro. She holds an MSc in software systems from the University of Aalborg and a degree in applied mathematics from the Technical University of Lisbon. Her main field of research is tourism network analysis and the study of strategic positioning with a multi-disciplinary approach. She has been involved in cross-group research (at both national and international levels), has participated in several conferences, and is author and co-author of international publications.

Grace Wen Pan is head of travel and leisure research at ACNielson China and an adjunct professor in the department of travel, leisure, sport and hotel management at Griffith University, Australia. She has been actively involved in China's outbound travel research for nearly ten years, and has conducted a number of research projects. She received her Bachelor's degree in Economics from Shanghai International Studies University, Master of Business (Research) from Queensland University of Technology, and Doctor of Philosophy in tourism markets from Griffith University, Australia.

Kathryn Pavlovich's research places collaboration at the core of inquiry. These studies include a ten-year ethnographic inquiry into network transformation, regional development through partnerships, and more recently a study on the role of intuitive knowledge in decision-making. Kathryn is an Associate Professor at the Waikato Management School, University of Waikato, New Zealand.

Christof Pforr's main research interests include tourism policy and planning, sustainable (tourism) development, coastal tourism, geotourism and ecotourism as well as policy analysis with a focus on policy processes and policy networks. Christof is a Senior Lecturer at the School of Management, Curtin Business School, Curtin University of Technology, Western Australia.

Ian Wilkinson's current work focuses on the development and management of inter-firm relations and networks in domestic and international business and the dynamics of markets and industrial networks, including an interest in complexity. Ian is a Professor at the School of Marketing, University of New South Wales.

Chapter 1
Introduction

We live in a networked world. The concept of a network of friends, of businesses or indeed of computers is pervasive in our conversations, newspaper articles or business plans. For many, the increasing importance of innovation and adaptation to turbulent environments is changing the nature of interaction with other organisations and as a response we increasingly encounter more networked inter-organisational relationships such as alliances, partnerships, clusters and communities of practice. These organisational forms often involve interaction between numerous individual organisations such that the flows of information and resources between them are complex. As a result these networks of organisations are becoming a dominant organisational form in the 21st century (Cravens & Piercy, 1994).

For many business sectors, the development of networks of organisations may be new or novel. For example, 'Just In Time' manufacture, which requires a network of suppliers working together, has been in place since the early 1980s (Huson & Nanda, 1995). In comparison, tourism has always been a networked industry and the usual description of tourism as a fragmented and geographically dispersed industry belies a pervasive set of business and personal relationships between companies and managers in businesses such as national tourism offices, hotels, attractions, transport, tours, travel agents and restaurants. It is this network of relationships that allows the tourism industry to deliver its product and to overcome the problems of fragmentation. Therefore it can be argued that the tourism industry provides the ideal context for study of networks.

The network concept is based around relationships between entities such as organisations or people (termed nodes), and the properties of networks studied by researchers relate to the structure of these relationships. The study of networks may be considered to have a number of paradigmatic characteristics (Wellman, 1988: 82) focusing on:

- Structural advantages and constraints on behaviour.
- The discovery of groups through their relationships rather than a priori allocation to categories.

1

- The overall structure of multiple relationships in a group rather than that between a particular pair of alters (in the language of network analysis, a particular node is identified as 'ego' and those nodes that ego has relationships with are termed 'alters').

One consequence of this approach is that it makes problematic the classical economic concept of a market as a homogeneous collection of identical suppliers and buyers. Instead, studying networks presupposes that the individuals do not act in isolation and with perfect information, but that the behaviour of individuals is profoundly affected by the pattern of relations that they may (proactively) develop. In studying networks the focus therefore is on relations rather than attributes, and on structured patterns of interaction rather than isolated individual actors. A second implication is that the fundamental basis for the study of networks is different from other areas which study the attributes of people or organisations. Instead, network analysis studies relationships (Knoke & Kuklinski, 1991).

Definition of a Network and Network Analysis

Originally, the concept of a network was a metaphor for the complex interactions between people in the community. However, with the development of quantitative approaches the concept of a network became formalised and related to mathematical theory. In graph theory a network is a:

> finite set of points linked, or partly linked, by a set of lines (called arcs) ... called a *net*, there being no restriction on the number of lines linking any pair of points or on the direction of those lines. A *relation* is a restricted sort of net in which there can only be one line linking one point to another in the same direction, i.e. there are no parallel arcs. (Mitchell, 1969: 2–3)

Transferred into sociology, a network is defined as a specific type of relation (ties) linking defined sets of persons, objects or events (Mitchell, 1969), and the sets of persons, objects or events on which a network is defined are called actors or nodes. Thus a network consists of a set of nodes, and ties representing some relationship between the nodes. Today, there are many definitions of a network but as pointed out by Jarillo (1988: 31), many have been developed by applying this basic definition to new areas such as the study of organisations where, for example, Gamm (1981) defines a network as a system or a field comprised of organisations and inter-organisational relationships.

Given this definition of a network, network analysis (or social network analysis) is an approach and set of techniques used to study the exchange of resources among actors such as individuals, groups, or organisations (Haythornthwaite, 1996). Because of this focus on relationships, the techniques used to analyse networks differ substantially from mainstream statistical methods that demand independent units of analysis. Network analysis therefore uses a set of integrated techniques to draw the patterns of relations among actors and to analyse their structure. The analysis is conducted by collecting relational data and organising it into a matrix and calculating various parameters such as density or centrality.

Network analysis has increased in popularity through the 1990s as an analytical framework, encouraged by the emergence of theories of society that emphasise relationships and integration. This is due in part to the effects of globalisation, which encourages alliances and linkages across organisations and nations, and to the greatly enhanced ease in communications encouraged by the wide diffusion of information technologies. In business and economics, network analysis represents a new organisational paradigm, drawing upon the competencies-based theories of the firm, where relationships shape and constrain organisational performance.

Within the tourism literature, the use of the concept of a network appears logical and delivers a number of useful outcomes for the analysis of tourism destinations and organisations. Tourism is a networked industry where loose clusters of organisations within a destination – as well as networks of cooperative and competitive organisations linking destinations – cooperate and compete in dynamic evolution. The concept of a network and the techniques of network analysis provide a means of conceptualising, visualising and analysing these complex sets of relationships. It provides a method for simplifying and communicating these relationships and so can be useful in promoting effective collaboration within destinations. It allows the identification of critical junctures in destination networks that cross functional, hierarchical or geographic boundaries, so ensuring integration within groups following strategic destination restructuring initiatives.

The aim of this book is to review the contribution of network analysis to the understanding of tourism destinations and organisations. We aim to provide an introduction to the use of quantitative network analysis for tourism and to provide some tourism applications of recent developments in network thinking derived from the physical and mathematical sciences. In working towards the achievement of these aims, we have reviewed the use of network analysis in tourism and found that the primary approach used to study destination networks is qualitative in nature. In this qualitative approach, the emphasis is on analysis using thick description where network diagrams, if used, are illustrative and display the relationships between pre-identified groups, rather than individual organisations or stakeholders. In comparison, much network analysis research outside tourism adopts quantitative methods where the emphasis is on collecting data concerning relationships between entities. These are mapped using mathematical techniques with results displayed visually in network diagrams and network attributes quantitatively measured.

This qualitative/quantitative divide echoes the qualitative–quantitative debate encountered in tourism and other fields of study (Davies, 2003; Walle, 1997). Outside tourism, this debate may be seen by comparing the inter-organisational network paradigm (Borgatti & Foster, 2003; Podolny & Page, 1998) with the policy network research tradition that emphasises qualitative and ethnographic methods (Rhodes, 2002). In policy network research, the focus is on the dynamic processes of policy-making, implementation and action derived from a view that the important focus for research is the individual. From this perspective, the quantitative approach to network analysis is seen as positivist and ignoring the changing nature of relationships with substantial methodological issues. A more balanced perspective is provided by Dredge (2005) who provides a framework for analysis of tourism policy networks that embeds the dynamic processes of policy-making within a structural network. From this perspective, the quantitative network approach used in this paper provides information on structural properties of the network as a whole that supplements the study of the relationships between individuals. A second differentiating characteristic of the quantitative social network approach is that it does not a priori define groups and structures within the destination. Instead, the aggregate network of relationships between actors in the network is used to define a group, cluster or clique. As Monge (1987: 242) writes, 'groups emerge by being densely connected regions of the network'.

Which is the better approach? Perhaps, when beginning this book, the authors may have been biased towards quantitative network analysis. However, the journey involved in producing a book such as this requires an understanding of the perspectives of many different authors, and it is clear that no single approach to the study of tourism networks can provide all the answers. The book is structured to reflect this debate and is offered to readers for them to choose the best approach, or indeed perhaps to chart a new approach that blends these two approaches together.

We have written the book to provide core ideas of network analysis and tourism, and have invited contributions from several specialists to augment and extend our thinking. As noted above, the qualitative/quantitative categorisation provides the basis for the structure of this book, effectively providing four sections – introductions, qualitative approaches to network analysis, quantitative approaches, and a concluding chapter.

The introductory chapter provides an overview of network analysis for tourism. It is followed by two chapters that provide firstly a history of the network concept in the social sciences and secondly an examination of the use of the network concept in the tourism literature.

The second section of the book reviews qualitative approaches to network analysis and tourism. Chapter 4 by Ian Wilkinson and Roger March provides a managerial application of network research in tourism and an example of how network analysis as a conceptual tool can be used by tourism managers to evaluate the effectiveness of their business-to-business relationships and partnerships. The chapter reports on an Australian Sustainable Tourism Cooperative Research Centre project aimed at developing a best-practice model for the efficient monitoring and organisation of relationships between tourism stakeholders in a regional tourism destination. In Chapter 5, Chris Cooper examines the management of knowledge in tourism destinations from a network perspective. Here knowledge is seen as a resource shared amongst stakeholders whose 'value' is determined in part by its distribution within the destination. The chapter develops a framework for knowledge management in a tourism destination and examines policy implications. Chapter 6 by Dianne Dredge and Christof Pforr examines the development of tourism networks as a new organisational form. The chapter asks if these new networked approaches are more efficient and effective in producing tourism public policy than the more centralised and bureaucratic approaches and if networks promote better tourism governance. In Chapter 7, Kathryn Pavlovich continues the discussion on network governance and network leadership in a case study set within an 'icon' tourism destination in New Zealand, the Waitomo Caves. The chapter examines the evolution of networks in the destination over a period of a hundred years focusing on recent capacity building and the development of knowledge network over recent years. Carlos Costa, Zélia Breda, Rui Costa and Joana Miguéns in Chapter 8 examine whether networks and clusters can be used as an innovative means to support tourism enterprises. They have conducted an empirical study in Portugal, targeting sports and adventure tourism enterprises, mainly consisting of SMEs. They suggest that by cooperating in the form of geographical and product-based clusters, enterprises can function as dynamic and interesting innovative organisations. In Chapter 9 Grace Wen Pan examines the cross-cultural context of network development. She examines the development of partnering relationships between Australian inbound tour operators and Chinese travel agents in the inbound Chinese travel trade to Australia. The study demonstrates the complexity of network development across cultural boundaries and concludes that the process is embedded with cultural factors, such as guanxi, ethnic preferences and regional cultural differences. In Chapter 10 Giuseppe Marzano examines the process of branding destinations through a network of stakeholders. Here networks are seen not as simple collaborative efforts but also as the vehicle for the exercise of power.

In the third section of this book quantitative approaches to network analysis and tourism are examined. We begin this section in Chapter 11 with a brief overview of formal network concepts and mathematical approaches. This is followed in Chapter 12 by an examination of network visualisation techniques, as one of the important advantages of network analysis is that output can include diagrams which help illustrate structural issues within destination networks. In the next two chapters we apply these quantitative methods to the analysis of tourism destinations. Chapter 13 places network methods within the broader context of complexity and chaos theories and goes on to present the study of two tourism destinations. It is shown how the quantitative approach can help in identifying the main structural characteristics of destination networks and how some of these measurements can be related to issues, such as collaboration and cooperation, which so far have been analysed only by using qualitative techniques. Chapter 14 analyses the technological counterpart of socio-economic systems: the Web space, and proposes the usage of the outcomes of this investigation as indicators to assess both technological and social conditions in a destination. This chapter closes with a consideration of numerical simulation methods. Their usage, it is shown, can prove very effective and useful in analysing special situations, in forecasting future scenarios and in providing destination managers with tools to improve their capabilities of adaptation and reaction to events.

In the final chapter of the book we synthesise the various approaches to network analysis and its application for tourism researchers and provide a discussion of future research opportunities and agendas. The study of tourism networks and the use of formal network analysis techniques have much to offer tourism researchers and we hope that this book will stimulate further development of network thinking. In particular we feel that tourism provides a rich context for research that will allow new theoretical developments of the concept to emerge.

The authors would like to thank a number of people who have helped and supported us in our work, in particular the specialist authors of the chapters in this book – their contribution has considerably enhanced the coverage of network applications to tourism. We are grateful to Dallen Timothy who provided constructive commentary on the manuscript. Noel Scott would like to thank his wife Trisha and family for their support during the last two years and would like to dedicate this book to the memory of his mother Jean Scott who died in February 2007. Rodolfo Baggio wishes to acknowledge Magda Antonioli Corigliano and the colleagues at the Master in Economics and Tourism, Bocconi University for their support and the fruitful discussions, and Valeria Tallinucci and Carla Catastini for the invaluable help in collecting much of the data used for the analyses presented here. Finally, without the patience and the assistance of his family, little of this book would have seen the light. Chris Cooper is grateful to Amy Cooper for putting the final manuscript together and to his two colleague authors – in part this book saw its genesis in fruitful discussions with Rodolfo in Milan and the enthusiastic adoption of the concept with Noel in Australia.

Chapter 2

The Historical Development of Network Theories

This chapter has the daunting task of introducing the reader to the history of network analysis. It is daunting as there is an extensive literature on the history of social network analysis with many papers including aspects of the topic (Shulman, 1976; Cook & Whitmeyer, 1992) and also with books containing some sections written on the topic (Chiesi, 2001; Scott, 2000; Wassermann & Faust, 1994; Wellman, 1988). This chapter is slightly different from these other works in that the authors set the use of network concepts and analysis in the social sciences within the broader literature of mathematics. For the purposes of this book, the authors believe that it is better to connect rather than sever these two disciplines. Thus, the historical development of the network concept is divided here into two broad schools of thought; one mathematically based and the other based in the study of the social sciences, with these two schools merging to some extent around the middle of the 20th century. This provides an introduction to the latter chapters in this book which have a quantitative and mathematical basis.

From a mathematical perspective (as well as in the visualisation of social networks used in early sociometry), a network may be represented by a diagram in which the various elements are represented by dots and the connections among them by lines that link pairs of dots. This diagram is called a graph, and the branch of mathematics known as graph theory constitutes the framework providing the formal language to describe such objects and their features.

Graph theory is one of the few disciplines with a definite birth date. As for many other fundamental branches of mathematics:

> ... the origins of graph theory are humble, even frivolous. Whereas many branches of mathematics were motivated by fundamental problems of calculation, motion, and measurement, the problems which led to the development of graph theory were often little more than puzzles, designed to test the ingenuity rather than to stimulate the imagination. But despite the apparent triviality of such puzzles, they

captured the interest of mathematicians, with the result that graph theory has become a subject rich in theoretical results of a surprising variety and depth. (Biggs *et al.*, 1976: 1)

The origin of graph theory is commonly attributed to the Swiss mathematician Leonhard Euler (1707–1783) and to his paper *Solutio problematis ad geometriam situs pertinentis* published in 1736. In it, Euler considered the now famous problem of the bridges of Königsberg. The people of Königsberg used to entertain themselves by trying to work out a route around the city crossing each of the seven bridges once and only once. All the attempts had always failed, so that many believed that the task was impossible (Biggs *et al.*, 1976). Euler proved this impossibility, giving also a simple criterion which determines whether or not there is a solution to the general problem of any number of bridges connecting to any other number of bridges connecting any number of areas.

Apart from the solution of this specific problem, the real importance of Euler's paper is in that it considers the object of study from an abstract point of view, giving significance to the structural characteristics more than to the purely geometrical ones. Euler's work also became the cornerstone of that discipline envisioned almost a century before by Leibniz, the *geometria situs*, the branch of mathematics known today as topology.

A number of important papers were published on this topic in the second part of the 18th and in the 19th centuries (Cauchy, Kirchoff, Hamilton, Poincaré, to quote just the most famous authors), and a formal setting of these theories was developed 200 years after the Königsberg Bridges paper. In 1936, the German mathematician Dénes König (1884–1944) published in Leipzig the first systematic study of what he called *graphs* in his *Theorie der endlichen und unendlichen Graphen*.

In the early 20th century, the ideas and techniques developed for the study of these abstract objects were applied in a completely different discipline – sociology. Realising that a group of individuals can be represented by enumerating the actors in the group and their mutual relationships, sociologists and anthropologists started using graph theory and methods to describe and analyse the patterns of social relations (Freeman, 2004; Wassermann & Faust, 1994). Jacob Moreno (1934) was one of a number of 'gestalt' psychologists in the United States who examined the structural patterning of thoughts and perceptions. He developed the use of sociograms (a diagram of points and lines used to represent relations among persons) to identify the structure of relationships around a person, group, or organisation and hence to study how these configurations affected beliefs or behaviours, and founded the journal 'Sociometry' in 1937 (Scott, 2000: 9). Today, the term *social network analysis* has superseded the earlier *sociometry*; however, both refer to the analysis of social networks in part utilising graphical methods. Friendships among individuals, business relationships between companies, and trade agreements among nations are all examples of networks which have been studied by using these techniques.

The origins of network thinking in social thought have been attributed by Grabher to:

> Simmel's (1890) fundamental distinction between 'groups' (defined by some membership criterion) and 'webs of affiliation' (linked through specific types of connections). By highlighting the critical role of the position of actors in 'webs of affiliation' he laid the foundations for social network analysis. (Grabher, 2006: 164)

From these early origins, the analysis of networks has expanded into a number of different lines of research. As Scott notes:

> A number of very diverse strands have shaped the development of present-day social network analysis. These strands have intersected with one another in a complex and fascinating history, sometimes fusing and other times diverging on to their separate paths. A clear lineage for the mainstream of social network analysis can, nevertheless, be constructed from this complex history. In this lineage there are three main traditions: the sociometric analysts, who worked on small groups and produced many technical advances with the methods of graph theory; the Harvard researchers of the 1930s, who explored patterns of interpersonal relations and the formation of 'cliques'; and the Manchester anthropologists, who built on both of these strands to investigate the structure of 'community' relations in tribal and village societies. These traditions were eventually brought together in the 1960s and 1970s, again at Harvard, when contemporary social network analysis was forged. (Scott, 2000: 7)

From the sociological and anthropological point of view, networks form part of the structural tradition. In this tradition researchers hypothesise that variations in the pattern of relationships surrounding social actors affect the behaviour of the actors and that correspondingly, people also consciously manipulate situations to create desired structures (Stokowski, 1992). Wellman writes that:

> The concern of structural analysts with the direct study of networks of concrete social relations connects strongly back to post-World War II developments in British social anthropology. Then as now,

anthropologists paid a good deal of attention to cultural systems of normative rights and duties that prescribe proper behaviour within such bounded groups as tribes, villages, and work units. (Wellman, 1988: 83)

In the 1950s a group of researchers associated with the Department of Social Anthropology at Manchester University were influenced by the work of Radcliffe-Brown and metaphors of a 'web' or 'network' of social relationships. In 1954 Barnes, one of the Manchester group, used the concept of 'the social network' to analyse the ties that cut across kinship groups and social classes in a Norwegian fishing village. Not only did the network concept help him to describe more accurately the social structure of the village, but he also found that it was more useful than normative concepts in explaining such key social processes as access to jobs and political activity. Soon after, the work of Bott, another member of the Manchester group, brought the network concept to the wider attention of social scientists (Bott, 1957). Bott developed the first distinct measure of network structure – 'knit', (now called 'density') – to show that densely knit, English extended families were more apt to contain married couples who did most things independently rather than jointly.

Slightly later, a group at Harvard University in the United States introduced two mathematical innovations: the development of algebraic models of groups using set theory, and multidimensional scaling. Multidimensional scaling is a mathematical technique for translating relationships into social distance and for mapping them in the social space (Scott, 2000). These innovations stimulated efforts to map interpersonal relations and to develop fine-grained methods for describing their patterns. Subsequently, epidemiologists and information scientists began conceiving of the diffusion of disease, information, etc as a social network phenomenon.

At about the same time that many structural analysts were developing ethnographic and quantitative approaches to studying social networks, others were analysing political processes as a result of ties of exchange and dependency between interest groups and nation-states. Researchers within this tradition have seldom used structural analytic tools or techniques. Grabher (2006) examines the interchange of ideas between economic geography and economic sociology and suggests that the social network analysis approach has been bypassed. Few see themselves as structural analysts but do want to know how patterns of ties in social systems allocate resources unevenly. Rhodes (1990) has provided an analysis of the development of policy network thinking and argues for an ethnographic approach to the study of policy networks (Rhodes, 2002) describing how actors create networks, thus rejecting the structuralist approach such as that proposed by Knoke (1980). Thatcher (1998) suggests the policy network literature has developed within a series of phases, from network being a metaphor to an overarching framework for analysis. Structural analysts have developed 'resource mobilisation' analyses to explain political behaviour. Policy network researchers have shown such behaviour to be due to structured vying for resources by interest groups, and not to reflect the aberrant cravings of a mob. Their work has emphasised how patterns of links between interest groups structure coalitions, cleavages, and competitive relations and how direct and indirect ties differentially link individuals and groups to resources.

The study of inter-organisational relationships is another key area for research today, and again the area has been studied from a dialectic perspective (Zeitz, 1980).

The use of network analysis in other disciplines has slowly grown. Wilkinson (2001) discusses the use of network thinking in marketing and notes the use of network analysis methods by Iacobucci and Hopkins (1992) in marketing channel analysis, but suggests that researchers have made limited use of them to date (Arabie & Wind, 1994). Within the marketing literature the study of distribution channels has been strongly influenced by the industrial marketing and purchasing group in Europe in the 1970s and by concepts of relationship marketing.

Today network analysis provides an analytical framework for the discussion of theories of society and globalisation encouraged by a tendency towards alliances and linkages across organisations (Pavlovich, 2001; Thrift, 1996). In business and economics, network analysis represents a new organisational paradigm, based upon the competencies-based theories of the firm, where relationships shape and constrain organisational performance (Tremblay, 1998; Welch *et al.*, 1998). This theory argues that organisations evolve according to the capabilities they can leverage from the external environment. In other words, a firm's performance is not only dependent upon the resources of the firm itself, but also upon those of other firms and the nature of their relationships (Wilkinson & Young, 2002). This system is a viewed as a network comprising an architecture of nodes and interconnected relationships where the network structure is strongly correlated to function (Watts & Strogatz, 1998).

Recent Developments in Network Analysis

The most recent development in the study of networks is also derived in part from mathematics. In the early 1960s the Hungarian mathematicians Paul Erdõs and Alfréd Rényi (1959; 1960; 1961) published a number of seminal papers on *random graphs*. The problem they addressed was a fundamental question in the study of graphs, networks and interconnection phenomena: *how do these objects form? And how do they evolve over time?* The approach used was statistical and probabilistic and the model they developed, the Erdõs-Rényi (ER) model, has since become a standard model with the capacity to explain many of the characteristics of the networks encountered in the real world.

In the last years of the 1990s the theories developed by mathematicians to understand networks have been applied to the Internet. The internet revolution has had a tremendous impact on almost all aspects of modern life and has provided a huge mass of network data for researchers to analyse using sophisticated mathematical techniques. In particular, an eclectic and interdisciplinary group of researchers have provided evidence that the ER model was simply a crude approximation of only a special class of networks, and that many of the networks found in the real world, from the technological, the physical, the biological or the social worlds, exhibited characteristics and properties of a diverse nature. Physicists, mathematicians, computer scientists, biologists, economists, and sociologists are all equally contributing to the growth of the knowledge in this field. (Watts, 2004) indicates that three influential papers typify this new approach:

- *Collective dynamics of 'small world' networks*, by Watts and Strogatz (1998).
- *On power-law relationships of the internet topology* by Faloutsos *et al.* (1999).
- *Emergence of scaling in random networks* by Barabási and Albert (1999).

The results of this vast amount of work have reinforced the idea that the collective properties of dynamic systems composed of a large number of interconnected parts are strongly influenced by the topology of the underlying network (see the bulky reviews by Boccaletti *et al.* (2006), Watts (2004), Newman (2003), Albert & Barabási (2002) and Dorogovtsev & Mendes (2002). It is fair to say that the implications of this work for social networks are still being appreciated.

The study of networks extends across physical, social, technological and biological domains and provides an active area for research. It has generated a number of specialised journals (*Social Networks* and *Connections*).

The concept of networks seems to have become a new paradigm in which it is possible to describe a wide variety of systems and their dynamical behaviour. This is not too surprising. In a sense, the whole history of science can be seen as the quest for the explanation of the relationships among the elements that form a system. Physics, biology, mathematics and others have pursued these objectives since their pre-scientific stages. Indeed the idea that the relationships give a certain shape to a system and affect many of its basic characteristics and functions is examined in Euclid's Elements (written about 320 BC) in which two books are dedicated to the discussion the general theory of proportion and to the similarity of figures and the transformation of areas. The Platonic and Aristotelian theory of forms also provides philosophical support to these ideas.

In this chapter, we have set the development of network theory within its mathematical context in order to draw attention to these new interdisciplinary developments in network thinking.

Chapter 3
The Network Concept and Tourism

Is network analysis suitable for the study of tourism? Based on the discussion in this chapter the authors consider that it is suitable and in fact that tourism is a network industry par excellence. Networks in tourism may be more important than in other areas of the economy of many countries. For example, in an Australian study of 1500 enterprises across all sectors of the economy, Bickerdyke (1996) found that networks were prevalent in the service sector and that the largest number of networked firms in the service sector was involved in tourism. In fact, tourism has been defined as a system where interdependence is essential (Bjork & Virtanen, 2005) and collaboration and cooperation between different organisations within a tourism destination creates the tourism product (Pechlaner *et al.*, 2002; Tinsley & Lynch, 2001). Buhalis (2000) has stated that most destinations consist of networks of tourism suppliers and that the benefits of such networks are more profitable tourism destinations (Morrison *et al.*, 2004). Lovelock (2001) discusses the importance of inter-organisational relationships, collaboration and cooperation. A network approach to sustainability is necessary within an industry such as tourism, where a relatively large number of small actors with few resources cannot pursue sustainable development in isolation (Halme, 2001). Networks, whether based on informal local alliances, formal partnership agreements, not-for-profit local, regional or national tourism organisations or other governance structures, help to compensate for the fragmented nature of tourism.

The fragmented nature of the tourism sector has often been discussed in the academic literature (Leiper, 1990; Palmer & Bejou, 1995; Wang & Fesenmaier, 2007). Tourism is seen as geographically dispersed in often remote areas distant from source markets, consisting of small independent businesses with a high staff turnover operating in a turbulent business environment. In such situations the survival of operators depends in part on collective action (Dollinger, 1990) and thus the emergence of network structures is in the collective interest of tourism operators. Indeed, tourism's organisation in a country can be considered as a series of hierarchical networks (Pearce, 1996). Thus networks are a logical response to the context that tourism provides to business managers, and network theory may therefore help to understand the collective nature of organisational action, constraint and coordination within tourism.

A second reason for collective action in tourism is that many of the main resources of a tourism destination that are used jointly to attract tourists are community 'owned'. These may be physical resources such as beaches, lakes, scenic outlooks and national parks; built resources such as museums, art galleries and heritage buildings; or intangible resources such as destination brands or the reputation for friendliness of local people. Such collective action does not necessarily require a network organisation but, in a situation with a general lack of resources and where decisions related to tourism are not often seen within the government mandate, the response is often a network of interested stakeholders.

There are a number of purposes for networks within tourism and Morrison *et al.* (2004) identify three types of network stakeholder. Networks of independent commercial operators provide an opportunity to influence planning, collect information and gain commercial advantage as well as a mechanism for leveraging resources such as marketing and business development activities, and to obtain public sector grant funding. Alford (1998), for example, focuses on how regional tourist boards seek to establish a market position, and how they benefit from networking with other sectors of the industry. Public/private networks may be used by government to stimulate economic development. Networks of tourism academics may provide information regarding changing business environment conditions, market research on customers or ideas for new business opportunities. In the absence of dedicated research and development facilities in many countries, academic networks may provide a resource that maximises tourism research funding. Certainly networks have become an active area of research for academics, with the ATLAS conference in 2004 having as its topic '*Networking and Partnerships in Destination Development and Management*'.

Networks also provide a practical business benefit. They provide a mechanism for passing customers from one organisation to another, to the benefit of those organisations, and they provide the customer with a comprehensive tourism experience (Curran *et al.*, 1993). However, practically there is a limit to the number of these relationships, as managing relationships takes time and effort (Hislop, 2005). Grangsjo (2003) has found that the majority of tourism operators are involved in business relationships with the source of their business, and their type of business determines the nature of their contacts and networks. Gibson *et al.* (2005) list a number of benefits of tourism networks such as learning, increased business activity and community development.

Networks are found within particular types of operators as well as across tourism destinations. For example, in a study of events in Australia, Stokes (2006) identified networking of stakeholders where operators share similar issues and problems to that of tourism. Similarly in a hotel context, part of the reason for networks developing is to obtain information in a complex dynamic environment. Ingram discusses the importance of a network in the hotel sector where

> competing managers are embedded in a cohesive network of friendships (i.e., one with many friendships among competitors), since cohesion facilitates the verification of information culled from the network, eliminates the structural holes faced by customers, and facilitates the normative control of competitors. (Ingram & Roberts, 2000: 387)

In the remainder of this chapter the areas in which the network concept has been applied in the tourism literature are discussed.

Networks, Collaboration and Trust

The first academic concepts examined here that overlap with the study of networks are those of collaboration and trust. Collaboration can be defined as 'a process of joint decision-making among key stakeholders of a problem domain about the future of that domain' (Gray, 1989: 227). As an example, the domain of interest for a tourism destination may be tourism planning, tourism marketing or other activity. A key reason for the interest in collaboration and networks in tourism development is the idea that tourist destinations can gain competitive advantage by bringing together the knowledge, expertise and other resources of their stakeholders (Kotler *et* al.,1993). Thus the concept of collaboration provides a reason for a network to exist.

Collaboration involves exchanging information, altering activities, sharing resources and enhancing the capacity of another for mutual benefit and to achieve a common purpose (Huxham, 1996: 28). According to Getz and Jamal (1994: 155), collaboration is a process of joint decision-making among autonomous and key stakeholders of an inter-organisational domain to resolve problems of the domain and/or to manage issues related to the domain. (Selin, 1993; 1991) in his work on collaboration indicates that collaboration works through networks.

Indeed, many researchers claim that the broadly based ownership of tourism policies can bring democratic empowerment and equity, operational advantages, and an enhanced tourism product (Jamal & Getz, 1995;

Joppe, 1996; Murphy, 1985; Timothy, 1998). Araujo and Bramwell (2000) have concluded that the network of stakeholders involved in collaborative planning for Brazil's Costa Dourada Project included a relatively broad array of environmental and other interests and also provided some possibility that varied issues of sustainable development would feature in deliberations about the project's policies.

Trust can be defined, first of all, as a state of favourable expectation regarding other people's actions and intentions (Möllering, 2001) and can be considered based on the law or a moral imperative (Hjalager, 2000). Trust can be a basis for individual risk-taking behaviour, cooperation, reduced social complexity, order, and social capital (Sztompka, 1999). At the organisational level, actors tend to create stable relationships with trusted partners, and these stable ties accumulate, over time, into a network that provides network members with valuable information about future alliance partners. Saxena (2006) provides an examination of the importance of trust in developing into organisational or community links that underpin tourism in the Peak District.

Networks in Marketing

The concept of networks in the general marketing literature is extensive and has begun to be employed in tourism. In the marketing literature Webster and Morrison (2004) write that network theories have been applied to word-of-mouth communication, relationship marketing, information acquisition, and the diffusion and adoption of new products and services. Arabie (1994: 270) writes that 'social networks and their patterns of relationships are a fundamental fact of market behaviour and can and have been used effectively as a basis for marketing strategies'.

In tourism, the idea that firms form destination marketing alliances is a common area for study (Palmer & Bejou, 1995), and Blumberg (2004) has examined cooperative networks in destination marketing in New Zealand. In her study, Blumberg found that one of the main challenges for a tourism destination management organisation was the organisation of cooperative networks to gain industry support for the destination's marketing activities. In an interesting study of destination marketing, Grangsjo (2006) has examined the balancing act between competition and cooperation that must be faced in marketing networks in tourism. She found that such networks encouraged and were supported by social capital developed through trust, communication and time spent together.

Communication in Tourism Destination Networks

In a growing body of research, researchers argue that organisations that are able to transfer knowledge effectively from one organisational unit to another are more productive than organisations that are less capable of knowledge transfer, e.g. Argote *et al.* (1990), Baum (1998), Cegarra-Navarro, (2005). Knowledge transfer is the process through which one network member is affected by the experience of another (Argote & Ingram, 2000).

In the broader literature, communication, knowledge development and knowledge management are topics that are commonly examined from a network perspective. The consequences of networks for information flow include information needs, exposure, legitimation, routes, and opportunities (Haythornthwaite, 1996). Indeed, Monge and Contractor (1999) have discussed ten families of theories and their respective theoretical mechanisms that have been used to explain the emergence, maintenance, and dissolution of communication networks in organisational research. These are:

- Theories of self-interest (social capital theory and transaction cost economics).
- Theories of mutual self-interest and collective action.
- Exchange and dependency theories (social exchange, resource dependency, and network organisational forms).
- Contagion theories, (social information processing, social cognitive theory, institutional theory, structural theory of action).
- Cognitive theories (semantic networks, knowledge structures, cognitive social structures, cognitive consistency).
- Theories of homophily (social comparison theory, social identity theory).
- Theories of proximity (physical and electronic propinquity).
- Uncertainty reduction and contingency theories.
- Social support theories.
- Evolutionary theories.

The importance of networks for learning is only slowly entering the tourism literature, although Cooper and Scott (2005) have examined the importance of networks for knowledge dissemination. Saxena (2005) has found that relational frameworks are the key sources of 'learning' for a region and has applied this idea as a way of improving the marketing of a region in the face of an increasingly changing external market environment. Tremblay makes a similar comment regarding learning by suggesting that in the context of high uncertainty tourism managers survive:

through the establishment of network linkages encompassing information exchanges, trust-building communication network channels, quasi-integration, and joint planning. This kind of network alliance has many advantages; e.g. exploitation of scale and scope of economics, common supply-side coordination, technological and physical assets, coordinating complementary assets. (Tremblay 1998: 849)

Another key area for the study of inter-organisational learning in tourism is the area of sustainable development (Halme, 2001).

Pavlovich (2003) indicates that 'internal ties' can further be used to build a portfolio of interconnections for knowledge building within destination networks. Saxena (2005) discusses how relationships provide relational capital that influences the degree of learning by individuals in a region. Morrison *et al.* (2004) summarise the function of international tourism networks for learning and exchange of knowledge. Relational frameworks thus emerge as the key sources of 'learning' and successfully marketing a region in the face of an increasingly changing external market environment (Saxena, 2005).

Network Forms of Governance for Tourism

The term *governance* has traditionally been defined very broadly as a 'mode of organising transactions' (Williamson & Ouchi, 1981). A more precise delineation of the concept is offered by Palay (1984: 265) who defines it as 'a shorthand expression for the institutional framework in which contracts are initiated, negotiated, monitored, adapted, and terminated.' Kooiman (1993: 2) defines governance as the 'activities of social, political, and administrative actors that can be seen as purposeful efforts to guide, steer, control, or manage (sectors or facets of) societies'.

One form of governance that has received particular attention is the network form, which Jones *et al.* (1997) define as a select, persistent, and structured set of autonomous firms (as well as non-profit agencies) engaged in creating products or services based on implicit and open-ended contracts to adapt to environmental contingencies and to coordinate and safeguard exchanges. These contracts are socially, but not legally, binding. Network forms of governance are seen to be flexible, with the potential to respond quickly to changes in their environment, especially where a coordinator is able to leverage the expertise of member companies (Palmer, 1998). Network governance refers to self-organising, inter-organisational networks characterised by interdependence, resource-exchange, rules of the game, and significant autonomy from the state (Rhodes, 1997: 15). Network governance constitutes a 'distinct form of coordinating economic activity', which contrasts (and competes) with markets and hierarchies (Jones *et al.*, 1997). The related term 'network organisation' refers to inter-firm coordination that is characterised by organic or informal social systems, in contrast to bureaucratic structures within firms and formal contractual relationships between them (Jones *et al.*, 1997).

Network governance provides an alternative process where contracts are social rather than legal (Jones *et al.*, 1997; Pavlovich, 2001). Network governance occurs through informal social structures, characterised by a sense of common purpose and interests, which are self regulated. Jones *et al.* (1997) identify four conditions that may lead to the development of network governance. These are demand uncertainty and unstable supply, customised exchanges dominated by human skills, and competencies where there is dependence between partners derived from their blend of skills and knowledge. Complex tasks are achieved within tight deadlines and exchanges amongst stakeholders are frequent, with consideration for each other's needs and the formation of trusting relationships within the overall network architecture. These factors appear to apply to tourism destinations.

Networks and Social Capital

The area of networks and social capital is a very active one for academic research and a number of reviews have been developed (Adler & Kwon, 2002; Borgatti & Foster, 2003; Burt, 1997; Coleman, 1988; Kogut, 2000). Glover and Hemingway write that:

> Social capital lies in the persistent social ties that enable a group to constitute, maintain, and reproduce itself. Such ties establish reasonably clear boundaries through mutual recognition and obligation. They also allow group members potential access to resources held by others in the group, thus enabling an individual to increase financial capital through loans or information from another group member, expand embodied cultural or informational capital through connections to experts and connoisseurs, or enhance institutionalized cultural capital by ties to organisations that bestow valued credentials and honorifics. Social capital is not an individual possession, as are other types of capital, but is instead the collective possession of those who are connected by social ties. (Glover & Hemingway 2005: 389)

The conceptual basis for the concept of social capital is the fact of membership in what Bourdieu (1986: 248) describes as 'a durable network of more or less institutionalised relationships of mutual acquaintanceship and recognition'. One important precursor to the study of social capital was Granovetter's seminal work (1985) on the fundamental importance of people's embeddedness in economic exchange. Stemming from a simple but critical regularity in economic exchange 'most behavior is closely embedded in networks of interpersonal relations' (Granovetter, 1985: 504). Granovetter argues that this ubiquity of embeddedness accounts for much of the order (and disorder) that is found in both markets and in firms. Burt's (2000) comprehensive, millennial meta-analysis of work in networks and social capital identifies four key dimensions that define the character of inter-personal networks: network size, network density, network constraint, and network hierarchy.

In tourism, Hall (2004) suggests that networks and cluster relationships are a significant part of the development of intangible capital and are a major focal point for much contemporary discussion of regional development. For efficient destinations, shared values are important and Pavlovich (2003) observes that high network density – the number of ties linking stakeholders – forces organisations to conform because institutional values diffuse through the network. In other words, stakeholders assume their identity according to the network structure. Hall (1999: 274) argues that 'the predominance of narrow corporatist notions of collaboration and partnership in network structures may serve to undermine the development of the social capital required for sustainable development'.

Networks as Representations of Complex Systems

Complexity is a multidisciplinary concept derived from mathematics and physics that has been applied to a wide range of different fields, including those belonging to the world of economics and social sciences (Henrickson & McKelvey, 2002). There is little agreement on a formal definition of a complex system but it may be characterised as an ensemble of elements in which the single components maintain their diversity and individuality while interacting locally among themselves. The interaction process is autonomous and produces outcomes that, at a global scale, cannot be simply reconstructed with a (linear) composition of the individual contributions (Levin, 2003; Waldrop, 1992). Systems evolving and adapting to the dynamic environment in which they are embedded form the category known as *complex adaptive systems* (CAS). Traffic is a good example of CAS, as is the weather, the stock market, an ecosystem, the Internet, or a biological organism.

A network of actors connected by links representing their interactions is a useful representation of such a complex adaptive system. In studying networks, researchers from disciplines as diverse as physics, biology, sociology and psychology have applied common tools and techniques to understand their properties. The properties of networks do not apply to individual nodes (neurons, people, companies) but to the network as a whole; the network is the unit of analysis. Thus, network researchers study the density, size, or centrality of the network rather than the properties of individual nodes. For example, the density of human networks has been found to be related to the effectiveness of communication of ideas and innovations (Monge & Contractor, 2003) as well as the spread of disease during epidemics (Barabási & Bonabeau, 2003).

The literature of network research has shown that the structure (or topology) of a network is a predictable property that greatly affects its overall dynamical behaviour and explains a number of processes, from the diffusion of ideas to the robustness of technical networks to external attacks to the optimisation of the relationships among the network components. Network analysis techniques are a diagnostic method for collecting and analysing data about the patterns of relationships among networks, such as people in groups, or among organisations (Boccaletti *et al.*, 2006; Newman, 2003).

An important, although rather scarce, strand of literature has pointed out the necessity to apply the complex systems framework when studying tourism and tourism systems. In these works, the reductionist paradigm used in dividing a tourism system into components, assuming that the relationships between them are stable and static, is challenged as being unable to provide meaningful explanations for many outcomes ((Farrell & Oczkowski, 2002; Faulkner & Russell, 2001; McKercher, 1998; Russell, 2005). This line of research, mainly based on qualitative considerations, has also been reinforced and complemented by some quantitative assessments of the 'complexity' of a tourism system.

Although only recently introduced to the tourism literature, the physical network approach has already provided meaningful results. The combination of the main metrics describing a network with the available qualitative information on some tourism destinations has given insights into their structure, their characteristics and their functions. Moreover, some relationships between the topology of the relations network and the dynamical (historical) evolution of the systems have been identified (Baggio, 2007a; Baggio *et al.*, 2007; Baggio & Scott, 2007).

The above chapter has illustrated the diverse and extensive literature of networks in tourism. It is one that is embedded in theory and goes to the heart of many of its central discussions.

Part 1
Qualitative Approaches to Tourism Network Analysis

Chapter 4
Conceptual Tools for Evaluating Tourism Partnerships

IAN WILKINSON and ROGER MARCH

Introduction

Tourism is an industry characterised by high degrees of inter-dependency where the evolution or creation of collaborative relationships such as networks is more than a natural outcome; it is a managerial imperative. From the perspective of tourism marketing (which is the focus of this chapter), the high cost of tourism promotion and advertising, coupled with the economic benefits to be gained from increased tourism into regions, obliges governments of all levels to commit financially to the marketing of their destinations. The benefits of collaboration are no less significant for private tourism operators (Robson & Robson, 1996; Halme & Fadeeva, 2000; Björk & Virtanen, 2005; Novelli *et al.*, 2006). The imperative for cooperation is particularly true in regional areas where, arguably, the dependence on successful tourism marketing and management requires a far more collaborative approach by local communities, small businesses and individual operators than is required by their counterparts in the large commercial, metropolitan areas of Australia. This is due to the large numbers of small enterprises, the absence of experience in the tourism industry, and the lack of tourist icons that underpins much of tourism demand for major costal destinations in Australia for example.

This chapter specifically examines networks constructed to achieve marketing outcomes for individual actors, outcomes which are also contingent on and intertwined with marketing initiatives aimed at promoting the overall destination. Of these, collaborative marketing alliances between public and private sector organisations have become increasingly popular in regional Australia. Attracting more tourists can benefit not only the narrow financial objectives of tourism operators, but also the more diverse social objectives of the public sector. The benefits of collaboration in tourism have been investigated for two decades (Boivin, 1987; Gunn, 1988; Stevens, 1988). Local government organisations, for example, have three compelling reasons for being involved in the promotion of tourism: (1) increased tourism generates additional revenue for the local authority, by creating more local jobs and thus lowering unemployment. Increased expenditure on tourism may also improve the image of an area and encourage further tourism and non-tourism-related investment; (2) in regional areas, in particular, the private sector is unlikely to have the necessary financial (or managerial) resources to allocate for effective destination marketing; and (3) local government organisation is responsible for providing vital elements of the tourism experience, such as interpretation of cultural and historical sites, visitor information centres and upkeep of infrastructure such as signage, parks and gardens, street cleaning, car parks, and retail districts.

Palmer (1996) identifies three 'marketing practices' that local government associations (LGAs) could implement to enhance the quality of the tourist experience in the destination: firstly, regular tourism-related training for council staff not directly employed in tourism (e.g. car park attendants, street cleaners); secondly, regular consultation of tourism department/planning departments on matters of planning and conservation; and thirdly, regular consultation between the tourism officer(s) and technical services department on matters of car parking, street cleaning and public convenience provision. The work of Porter (1990, 1998) and others has contributed to the idea that networks represent coalitions of collective action, which are preconditions for innovation and community capacity building. Network theory assumes that 'relationships do not occur within a vacuum of dyadic ties, but rather in a network of influences, where a firm's stakeholders are likely to have direct relationships with one another' (Rowley, 1997: 890). In Australia, policy initiatives aimed at generating greater economic wealth through the creation of industry networks or clusters have typically ignored tourism. Roberts and Enright's (2004) overview of industry clustering in Australia makes no mention of tourism clusters, even though the authors examine, ironically, the Hunter region in detail. Tourism destinations lend themselves well to network analysis, since they comprise multiple suppliers from a range of businesses and sectors. It is the quality and complementarity of these suppliers and other tourist services available that will determine the overall appeal of the destination experience, just as the effectiveness of inter-organisational relationships largely determine the effectiveness and efficiency of, among other things, collaborative marketing activities.

At the regional level, the typical tourism network is the nexus between state tourism organisation, the regional tourism organisation, the local council(s) and tourism operators. This network works more efficiently and effectively in some regions than others. Organisations in a network can be regarded as economic actors, which are inter-related through a web of resources and activities. Value is created in a network by actors who perform and control activities that are based on control over critical resources, and include social content by developing relationships with each other through exchange processes (Hakansson & Johanson, 1992; Axelsson & Easton, 1992). Though critical resources can be physical, in the tourism industry they are mainly knowledge-intensive intangibles.

Though one of the key determinants of the sustainable growth of the tourism industry in regional areas is the development of effective networks and partnerships between actors, research on the topic has been sparse. One recent study by Dredge (2006a) analyses the historical development of the networks created by a local tourist association in the Lake Macquarie region of Australia. A key finding in that study was that private and public sector participants require an understanding of the power differentials between actors, and an awareness of the different opportunities for actors to participate in leadership of the industry. Such an understanding is important in determining strategies to encourage engagement and to harness the contributions of diverse local government and industry players. This chapter seeks to identify power differentials from a resource dependency perspective. Relationships can be seen as a dynamic, on-going, socially constructed and often negotiated process involving multiple actors. Verbole (2000) argues that in order to understand tourism in the rural (or regional) areas, the social actors (social networks, factions, and formally constituted groups as well as local councils) need to be identified and processes need to be investigated. By identifying various networks and other organisational practices it is possible to gain insights into decision-making for tourism promotion.

Networking in this chapter is defined in terms of social interactions between various actors. Leeuwis (1991: 13) defined actors' networks as '...flexible and changing sets of social relations between individual and institutional actors that involve material, social and symbolic change'. Business networks have their basis in social relations (Granovetter, 1985) and the 'creation of a regional network is not possible without social relationships between the actors that lead to a regional culture' (Lechner & Dowling, 1999: 312). Networks can be formally structured or can exist simply as informal agreements or arrangements between people and/or organisations, based on single or multi-purpose social relations.

Mitchell's (1969) classical sociological study identifies three different ways in which the content of social network links may be perceived – exchange, communication and social. The nature of such linkages exists on a continuum ranging from 'loose' linkages to coalitions and more lasting structural arrangements and relationships. Applying Mitchell's typology, three network links can be postulated from the perspective of a tourism enterprise operating in a regional destination. The first is exchange network, in which there are businesses and organisations with which the tourism operation has commercial transactions; the second is communication network, whereby organisations with non-trading links inform the operations of the tourism business, for example, LGAs, state government departments, consultants, and industry or sectoral associations. The third is social network of family, friends, and acquaintances of the business owner. These can be further split into the personal network, involving concrete contact with specific individuals, and the wider cultural dimension in which actors are immersed, namely, the values, attitudes and behaviour that significantly influence the nature of the relationships formed. Mandell (1999) identifies a continuum of collaborative efforts as follows:

- Linkages or interactive contacts between two or more actors.
- Intermittent coordination or mutual adjustment of the policies and procedures of two or more actors to accomplish some objective.
- Ad hoc or temporary task force activity among actors to accomplish a purpose or purposes.
- Permanent and/or regular coordination between two or more actors through a formal arrangement (e.g. a council or partnership) to engage in limited activity to achieve a purpose or purposes.
- A coalition where interdependent and strategic actions are taken, but where purposes are narrow in scope and all actions occur within the participant actors themselves or involve the mutually sequential or simultaneous activity of the participant actors.
- A collective or network structure where there is a broad mission and joint and strategically interdependent action. Such structural arrangements take on broad tasks that reach beyond the simultaneous actions of independently operating actors.

However, as Mandell, cited by Hall, cautions:

> because we as professionals are eager to achieve results, we often look for prescriptions or answers as to how to solve ongoing dilemmas …it is tempting for both academics and practitioners to try to develop a model of success that will fit this complex world. In this regard, the concepts of networks and network structures can easily become the next in line for those in the field to 'latch onto' and use wholesale. Although it may be tempting to do so, this 'one size fits all' type of modelling does not take into consideration the myriad of factors and

events that must be understood before these concepts can be of much use in the 'real world'. (Mandell 1999; Hall (1999: 277)

Introducing the Research Setting

State tourism regions in Australia are designated by their respective state governments. They are administrative regions rather than tourism regions, each with their own distinctive bundle of tourism experiences. While the administrative boundaries may be clear, from the perspective of tourism operators and especially tourists, the boundaries are much fuzzier. Operators are likely to be part of the value chains and complementary and competitive networks of tourist experience providers that cut across sub-regions and even states. From a tourist's perspective, administrative boundaries mean nothing; he or she cares only for what contributes or detracts from the tourism experience.

The proximity of the Hunter region to Sydney influences the number and types of tourists visiting the region, predominantly domestic and international overnight as well as day trip visitors. The north-eastern perimeter of the Hunter is a gateway from Sydney to the northern New South Wales coastal resorts, to inland New South Wales and further afield to Queensland. The development of direct domestic and international flights into Newcastle Airport is changing the psychic boundary of the region, creating opportunities and threats as the airport management seeks to tap into new market segments while, at the same time, opening up opportunities for destinations outside the Hunter.

The region is characterised by a great diversity of tourist experiences. These vary from wine tourism around Cessnock and Pokolbin, for which the region is well known, to nature-based experiences such as the Barrington Tops, horse riding and rearing (e.g. Scone), beach and coastal resorts (e.g. Port Stephens area), lakes and water sports (e.g. Lake Macquarie), national heritage and city experiences (e.g. Newcastle and cruises). This diversity, as we shall see, presents challenges in coordinating and integrating regional promotion campaigns and marketing strategies. This is further exacerbated by the division of the region into twelve LGAs, each with its own tourist administration and experiences.

A conundrum of the Hunter region is that while many stakeholders in the region not associated with wine complain that the overwhelming and dominating image of the region is 'wine', less than 15% of domestic visitors to the Hunter actually visit a winery (Tourism NSW, 2003). Indeed, it was the consumers' perception of the Hunter being about wine that led Port Stephens Tourism Association to formally withdraw from the Hunter

Tourist Region and align itself with the Mid North Coast tourism region. In contrast, the tourist area of Gloucester, located on the northern inland boundary of the region, and which includes part of the Barrington Tops, joined the Hunter Region because of the perceived closer tourist experience 'fit' with the Hunter Region.

In recent years, there has also been an influx of new players into the region, into tourism in particular and the service industry in general, where ease of entry has led to a growth in the number of small operators such as B&Bs, boutique wineries, restaurants and attractions (hot-air ballooning and alike). Large-scale investment in tourist-related facilities is taking place, including hotels and accommodation (e.g. Crown Plaza and The Vintage development in Cessnock, French Village in Maitland), resorts (e.g. Eaglereach Nature Resort) and major attractions (e.g. Hunter Valley Gardens and the Honeysuckle Development in Newcastle).

Ironically, as new entrants have appeared in the region, tourism demand has been falling. Visitation to the Hunter region has declined in recent years in the key domestic market. International visitation has been stagnant since 2000, while the number of domestic visitors has dropped from 2.8M in 2001 to 1.9M in 2005 and domestic day trips have fallen from 4.2M in 2003 to 3.7M in 2005 (Tourism NSW, 2006). The Hunter region is primarily a day trip or overnight destination, with the exception of the coastal areas around Port Stephens, which attract more annual holidaymakers.

The maturity and history of the region is reflected in the changing industry mix in the region. The industries that have dominated the region are related to farming, mining, steel production and shipping. Apart from wine, each of these industries is in decline for various reasons. Tourism is seen as an important potential source of future income, employment and economic growth, and this is not any easy transition for many communities and industries.

Methodology and Initial Findings

The Hunter Valley region was chosen by the authors after consultation with research managers at the state tourism organisation for New South Wales, Tourism NSW. This is one of Australia's most famous wine districts, located 160 kilometres northwest of Sydney and a major tourist destination for local and international visitors. This choice of the Hunter region was based on its perceived success as a regional tourist destination: infrastructure investment is relatively strong, and coordination amongst tourism stakeholders is regarded as effective and efficient, and in particular there was strong involvement of the private sector with the regional tourism organisation, the Hunter Regional Tourism Organisation. A list of key stakeholders in the local tourism industry was requested from and supplied by Tourism NSW.

A case-study approach was adopted. Networks are complex social organisms, comprising formal and informal actors, and only in-depth interviews with numerous individuals can adequately reveal the dynamics of relationships among actors. Interviews were conducted with fourteen respondents. Most interviews were one-on-one, although on several occasion two respondents were interviewed together. (For example, the general managers of two resorts and the head of the regional tourism organisation were interviewed together.) All interviews were taped (provided permission was received) and transcribed. The interviews proceeded in a series of three main waves in July and August 2004, starting with the original list and then including other types of actors. Interviews were conducted by the two authors. Snowballing techniques were used whereby at the completion of each interview, the respondents were asked to provide the researchers with further contacts who were likely to have different perspectives on the tourism industry, or play different roles that had not been included in the original list of contacts. The interviews included the following organisations and firms: resort operators, wineries, tourist attractions, local councils, state government agencies, local council tourism managers, airport authorities, trade and industry associations, government tourism agencies, non-tourism government agencies, wholesalers, and inbound tour operators. Access was also granted to the draft marketing plan for the Hunter Region prepared by Calais Consulting. This provided useful background statistics and also served as another source for identifying issues and improving the researchers' understanding.

Two major tourism industry associations operate in the region. Hunter Valley Wine Country Tourism (HVWCT) is an incorporated local tourist association that runs the major Visitors Information Centre in the Hunter Valley. It is an incorporated body with a membership base of over 550 businesses. HVWCT also produces regional guides for the tourist, conference and wedding markets. The regional tourism organisation (RTO) for the Hunter is the Hunter Regional Tourism Organisation (HRTO). This organisation works with operators on regional promotion and regional campaigns, and provides the conduit for operators and local government tourism associations to receive cooperative funding from Tourism NSW. Various sources of conflict and tension in relations were identified among those involved in the region. Such conflict is not necessarily bad as it reflects the complexity and ongoing adaptation of an industry to changing circumstances. What matters is whether such conflict can be managed productively and channelled into meaningful responses and learning, rather than into damaging turf wars and destructive responses.

In the following section, we introduce the research findings through the prism of four analytical tools that, we argue, have the potential for providing fresh insights for the optimisation of network relationships.

The Value Nets of the Main Types of Actors in the Regional Tourist Industry

There are several ways of classifying the main types of actors participating in the tourism value chain. For our purposes, the central actors are the tourism operators who, in the Hunter region, are mainly accommodation operators and wineries, but also event organisers and transport providers. There are those that provide support services to tourists, such as information services (e.g., media, guide books and websites, as well as advice from travel agents) and wholesalers, inbound tour operators and travel agents who sell packages and assist in tour planning.

Frontline tourist actors interact with various types of other actors in carrying out their activities. These interactions may be depicted in terms of their *value nets*. Value nets involve four generic interaction types in which a focal tourist actor is involved: *competitors* whose outputs reduce the value of the focal actor's output (other tourism actors, intra and interregional competitors, indirect competitors); *complementors* who enhance the value of the focal actor's outputs (other tourism actors, support services, government organisations, trade and industry organisations); *suppliers* (of staff, provisions, materials, technology, finance, services and other component inputs); and *customers* (tourists, and channel intermediaries linking a tourist operator with actual and potential tourists). The main actors in the tourist industry in the Hunter region were examined in terms of their value nets. See Figure 4.1 for example of this model using a tourism resort as the core actor.

The value net reveals two fundamental symmetries in the game of business. On the vertical dimension, customers and suppliers play symmetric roles. They are equal partners in creating value. But managers do not always recognise this symmetry. Brandenburger and Nalebuff (1996) argue strongly that supplier relations are no less important for businesses than customer relations. If we accept that a tourism business can enhance its competitiveness by adding value to its offering in a more cost-effective means than its competitors, then the development of stronger relations with suppliers is critical to that goal. Symmetry also exists on the horizontal

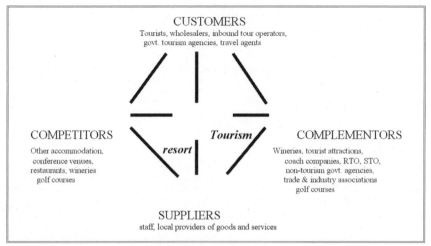

Figure 4.1 Value net: The example of a Hunter Valley tourism resort

dimension. However, this time the difference is that competitors and complementors play mirror-image roles. An actor would be the resort's complementor if customers value the resort's product more when they have the other actor's product than when they experience the resort's product alone. In other words, without the surrounding wineries, tourist attractions, restaurants, and golf courses the resort would be valued less. Conversely, an actor would be the resort's competitor if customers value the resort's product less when they experience or consume the other actor's product than when they have the resort's product alone. Obviously if a tourist couple are playing golf and lodging at a luxury golf course they are valuing the resort less. (Space limitations do not allow for a fuller explanation of the strategic implications of value nets in the tourism space.)

Partnership–Activity Matrix

The second means of analysing the interactions of actors in the Hunter region is by matching the types of partnerships identified in the field trips with the (mostly) marketing-related activities that these partnerships enacted. The main partnerships identified were: among LGAs; between operators and tourist associations; among neighbouring tourist associations; between leading tourist operators and local produce providers; among transport services within and to the region; with collaboration with external organisations; and partnerships with regional education and training providers (see Table 4.1). Several lessons emerge from these findings. First, operators

Table 4.1 Evaluation of relationship types

	Partnership Types						
	Among LGAs	*Operators & tourist associations*	*Neigh-bouring tourist associations*	*Leading tourist operators & local produce providers*	*Partnerships in transport services within & to the region*	*Collab-oration with external organi-sations*	*With regional education & training providers*
Symbiotic	✓			✓		✓	
Neutral							✓
Competitory	✓		✓		✓		
Predatory							

generate the most number of joint activities with local produce providers, which confirms the argument of Brandenburger and Nalebuff (1996) that suppliers are as important as customers. Secondly, partnerships serve multiple purposes, not simply to save costs on advertising costs; gathering market intelligence was a commonly cited aim of partnerships. Thirdly, LGAs displayed little inclination to engage in collaboration despite the obvious synergies that cooperation among neighbouring tourism regions offers in targeting the drive market, in engaging in joint advertising, and in formulating a more coherent positioning or bundling strategy for their combined geographical segment of the designated tourism region.

Ecological Approach to Classifying Partnerships

Fennell and Butler (2003) applied a human ecological approach when examining interactions in a tourism system by drawing upon the earlier work of Budowski (1976), who argued that relationships between tourism stake-holders could be located along a continuum from predatory to symbiotic. Relationships are predatory when one stakeholder exerts a high level of impact on a community or network. A competitory relationship is one in which competition exists for resources, whether they are tourists, natural resources, promotional funding, tourism operators (in the case of industry associations competing for association membership) or funds for tourism infrastructure development. Neutral relationships occur between stake-holders that have little or no impact on each other; finally, symbiotic relation-ships between stakeholders are those where the objectives of the stakeholders are achieved more efficiently by collaboration between stakeholders. Adapting this typology, we can categorise the partnerships in Table 4.2.

Table 4.2 Partnership-activity matrix

	Partnership Types						
	Among LGAs	Operators & tourist associations	Neighbouring tourist associations	Leading tourist operators & local produce providers	Partnerships in transport services within & to the region	Collaboration with external organisations	Partnerships with regional education & training providers
Activity Types							
Pooling expertise and resources to better access external funding			✓	✓			✓
Joint promotional activities		✓	✓	✓		✓	
Lobbying activities	✓				✓		✓
Support for local and regional events		✓		✓		✓	
Tourist product bundling			✓	✓			
Joint development of infrastructure					✓		
Joint gathering of market intelligence	✓	✓	✓	✓			
Wine and tourism marketing		✓		✓			

Resource Scarcity

Numerous sources of conflict and tension between organisations were identified across the Hunter region. We have conceptualised this conflict as competition for scarce resources, which we have identified as private sector and government funds, environmental resources and tourist segments. The stakeholders involved in conflicts are listed in Table 4.3. (For the sake of parsimony, each conflict was only allocated to one competing resource.) We

Table 4.3 Stakeholder conflicts, by scarce resource

Public sector funds	Private Sector Funds	Environmental resources	Tourist Segments
Region vs state and Sydney Community needs vs tourism demand Among local government areas Among tourist associations for members	Among tourist associations Large vs small operators Mature vs developing destinations	Vignerons vs agriculturists/tourism operators Tourism vs other industry development Local vs regional focus	Wine tourism vs non-wine tourism Competition among types of tourist experiences Geographic market focus vs customer segment focus Wine sales vs tourism services Large vs small operators Mature vs developing destinations

argue that awareness of the competition for scarce resources is essential for actors in tourism regions, particularly when the region is lacking substantial private sector investment and public funding. We offer three key implications for strategy: first, government bodies that fund tourism marketing should seek to reduce inter-organisational competition by incorporating competing organisations and regions into cooperative partnerships; secondly, regional tourism organisations could partially circumvent potential conflict for scarce resources by developing marketing strategies.

Limitations and Suggestions for Further Research

The participants interviewed for this project are not a representative sample of actors in the Hunter tourism region, although the interviews did generate a number of other potentially valuable individuals and organisations that could be interviewed for the proposed second stage of this project. Notwithstanding this limitation, the authors are confident that the main issues and characteristics of partnerships and inter-organisational networks in the region have been identified. These findings provide the basis for developing conceptual tools that tourism managers can apply when evaluating their present and future tourism partnerships.

A number of future research opportunities are suggested by this research. First, mapping and explaining the structural change and evolution of tourist regions in terms of local industry clusters would provide indicators for the development of future clusters in developing regions, or the reorganisation of existing clusters. Secondly, an audit of the structural integrity of existing tourism region networks, including the planning, conduct and performance of inter-organisational relationships, would generate further insights into the sustainability of such networks. Thirdly, research that investigates the links, strength and nature of inter-organisational ties to behaviour and performance of firms would further enhance the application of network theory by tourism managers.

Conclusion

The research detailed in this chapter was funded by the CRC for Sustainable Tourism, the aim of which was to develop a best-practice model for effective network organisation in regional tourism areas. This chapter has examined the nature and type of network relationships involving tourism-related stakeholders in the Hunter Region. The chapter also offered conceptual tools for analysing existing and potential partnerships at the individual organisation level. The ambition to develop an operationalisable framework for building effective and sustainable networks remains an ambition.

The reality is that tourism networks are organic entities that develop over time in response to environmental and organisational demands. Networks comprise a multitude of stakeholders that cross sectoral, administrative and geographical boundaries. A network does not have a core, realisable goal; in fact, a network does not have predetermined organisational structure and for that reason it cannot, by definition, be managed. Fennell and Butler (2003: 208) echo this perspective: 'tourism in destination regions or communities is rarely managed, even when it has been planned'. Kaltenborn (1996: 26) adds another dimension to our understanding of the problems that we face: 'tourism management and the commercial tourism industry represent two very different systems that operate on very different assumptions and requirements, but that often lay claim to the same resources'. The key here is competition for the same, usually limited, resources. Identifying these scarce resources and minimising the competition for them may assist in the development of sustainable tourism networks.

Chapter 5

Tourism Destination Networks and Knowledge Transfer

Introduction

Destinations are undergoing fundamental structural change driven by technology, changing systems of governance and the demands of the knowledge economy (Buhalis, 2000). This chapter examines the issues surrounding the process of knowledge transfer across tourism destination networks. The generation and transfer of tourism knowledge is essential for sustainable innovation at destinations and this in turn underpins competitiveness. Yet, despite the fact that knowledge is increasingly seen as the engine of economic growth in regions (OECD, 2001) this is a field that has received little attention in the tourism literature – in stark contrast to other areas such as research and development and regional planning. The network perspective of destinations championed by this book highlights the importance of shared knowledge. There is no doubt that the generation and use of new tourism knowledge for innovation and product development is critical for the competitiveness of destinations. In fact, despite the fact that researchers, consultants, the industry and government are constantly generating new tourism knowledge, destinations have been slow to harness that knowledge. This chapter outlines the concepts and processes of knowledge creation and dissemination within a regionally bounded network of destination organisations and concludes with the policy implications for tourism.

The Knowledge-based Economy

Tourism has been slow to recognise the significance of the knowledge-based economy. However, the new paradigm represented by this approach, and the tools of knowledge management have much to offer tourism. In the late 1990s, the knowledge-based economy emerged from the previous information age. There was a recognition that not only was knowledge more than information but also that it was a resource to be valued and managed. The knowledge economy can therefore be thought of as an economy directly based upon the production, distribution and use of knowledge. The knowledge-based economy has two important new features that demand a rethinking of our approach to both tourism destinations and tourism policy. These features include:

- Structural economic change, as the new tourism products and innovations brought about by technology develop. Technology, particularly the Internet, breaks down barriers to knowledge sharing through the use of such innovations as destination knowledge portals.
- The employment of highly skilled labour as a means of competitive advantage and long-run economic growth. This is an important issue for tourism where many of the sector's human resource practices mitigate against employment and retention of highly skilled employees. Indeed it could be argued that whilst most destinations used to compete on the basis of endowed factors such as natural beauty or climate, in the knowledge based economy it will be their personnel and levels of service that will act as competitive differentiation. In other words the creation of a new paradigm of 'knowledge commerce' will render traditional competitive measures – such as location – less important.

Types of Knowledge

It is important to understand the intrinsic nature of knowledge and the key types of knowledge that are available in tourism destination networks. The knowledge management literature provides useful insights for tourism in terms of concepts and definitions of knowledge. Knowledge can be thought of as actionable information, available in the right format, at the right time, and at the right place for decision-making. To quote Davenport and Prusak:

> Knowledge is a fluid mix of framed experience, values, contextual information, and expert insight that provides a framework for evaluating and incorporating new experiences and information. It originates and is applied in the minds of knowers. (Davenport and Prusak, 1998: 5)

Knowledge management classifies knowledge according to its ability to be *codified* and therefore communicated. For tourism this distinction is fundamental and goes a long way to explaining the failure of the tourism sector to adequately capitalise upon and manage knowledge. Polanyi (1966) provides possibly the most useful classification, distinguishing between two types of knowledge, tacit and explicit.

Tacit knowledge

Tacit knowledge is difficult to codify, difficult to communicate to others as information, and it is difficult to digitise. A good example of tacit knowledge would be the knowledge that is passed from a tourism destination marketing manager to an assistant in terms of designing and managing promotional literature. The majority of knowledge in tourism destination networks is tacit (for example, in tourism organisations and the entrepreneurial community). Yet despite the fact that estimates suggest that over 90% of an organisation's knowledge assets are tacit, this type of knowledge is often ignored. Effectively then, tacit knowledge is the practical knowledge needed to perform a task. Here, the link between tacit knowledge and destination competitiveness is clear. Tacit knowledge is possessed by the personnel of a tourism company, government ministry or destination management organisation, and it is this tacit knowledge that is difficult for competitors to access or replicate. In other words, the tacit knowledge of an organisation or destination is one of its core capabilities, as it distinguishes an organisation from its competitors and promotes strategic advantage.

However, the fact that it is individuals who possess tacit knowledge can create tensions with the employing organisation. This is because tacit knowledge is the basis of an individual's personal competitive advantage – particularly for, say, entrepreneurs, and so there is often a reluctance to share or communicate it. In addition, this tacit knowledge, by definition, cannot be questioned or discussed because it has not been communicated to the rest of the organisation. Nonetheless, despite the difficulty of characterising tacit knowledge, it can be evaluated because it has objective and tangible consequences such as profitability or customer satisfaction. In other words, the outputs of tacit knowledge can be tested for quality.

Explicit knowledge

In contrast to tacit knowledge, explicit knowledge is transferable and easy to codify and communicate. It is therefore usually the focus of an organisation's interest and is found in the form of documents, databases, files and other media. Explicit knowledge can be relatively easily formulated by means of symbols and can be digitised. It can therefore be relatively easily transferred and communicated to those that need it at the destination.

The implication of Polanyi's (1966) classification of knowledge is that the conversion of tacit to explicit knowledge is critical for tourism. Quite simply, the majority of knowledge available to a tourism destination is tacit knowledge. It is here that the knowledge management approach provides a significant benefit for tourism as it focuses upon the management of tacit and explicit knowledge to create organisational learning, innovation and sustainable competitive advantage for a destination (Cooper, 2006).

Knowledge as a Resource

As well as understanding the types of knowledge available to destination networks, it is also important to recognise the nature of knowledge as the resource to be transferred across the network. A range of distinctive characteristics of knowledge as a resource can be identified:

- Knowledge is difficult to own and control, and can have many owners – market research reports would be an example here.
- Tacit knowledge makes knowledge 'sticky' and difficult to turn into explicit, communicable information – a major issue in tourism.
- Knowledge delivers increasing returns. Unlike most resources, the more it is used the more benefits it delivers.
- Knowledge can be used without being consumed.
- Knowledge evolves and changes, as evidenced by the tourism sector's growing knowledge base of crisis and risk management, building upon a very low base prior to September 11th.
- The value of investing in knowledge is uncertain and difficult to predict – investment in knowledge is heavily front-ended, hence the need for governments to invest in the collection of data for national and regional tourism surveys, and the reluctance of the tourism private sector to invest in tourism research.
- Knowledge is generally context-specific, created in *communities of practice*. This is a major issue for tourism where there are two distinct communities of practice at a destination – the researchers who generate knowledge, and the sector itself that uses it.
- Knowledge is specific – it cannot be easily transferred from one activity to another.

Knowledge Management and Tourism Destinations

Elsewhere in this book, the nature and structure of tourism destinations as networks have been elaborated. Effectively, destinations can be thought of as loosely articulated networks of enterprises, governments and other organisations. Collectively, they have the overall goal of ensuring both the competitiveness and sustainability of the particular destination. This is potentially a problem when examining a knowledge management approach to destinations, as the majority of the literature and applications are concerned with an individual organisation. Of course, the focus on the individual organisation can be applied to tourism enterprises, destination management organisations and to government ministries and departments. However, if knowledge management is to be an effective tool for competitive destinations, then we also need to consider how it can be applied at the destination scale. In other words, current thinking in knowledge management needs to be expanded to embrace knowledge stocks and flows within heterogeneous networks of organisations at the destination. Here, Hislop *et al.* (1997) provide a solution by arguing that knowledge 'articulation' occurs in networks of organisations attempting to innovate and build upon knowledge. They identify two types of network:

(1) Micro-level networks within organisations where knowledge is created and is dominantly tacit and 'in-house'. This can be thought of as 'demand-side' knowledge creation, satisfying the organisational needs of new knowledge and is learning or innovation-centred.
(2) Macro-level, inter-organisational networks where knowledge is transferred around a network of organisations and tends therefore to be explicit in nature. This can be viewed as a 'supply-side' response to the need to distribute and transfer knowledge.

Hislop *et al.*'s (1997) notion of knowledge 'articulation' involves the gradual conversion of tacit knowledge at the individual organisation level into explicit knowledge which is transmitted through the wider network of organisations by the usual processes of knowledge management. In this way useful knowledge is widely dispersed across a network to boost competitiveness, and the analogy with tourism destinations is clear. Knowledge is created by individuals, validated by communities and used by individual organisations or networks within destinations to innovate. In tourism, knowledge management embraces both levels of aggregation; i.e. both the individual organisation as well as networks of organisations. At the destination level, the networks of organisations can be either destination or sector-specific and either will facilitate the knowledge management process:

- 'Traded interactions' where knowledge sharing is facilitated by members of the supply chain or by trade organisations for industry sectors.
- 'Untraded interactions' where knowledge sharing is facilitated by civic activities via destination management organisations for the destination itself.

In both cases, the facilitation of knowledge sharing demands high degrees of trust and the often competitive environment of destinations can work against this. It also raises the question as to whether the knowledge is specific to the destination or more general. 'Local' forms of knowledge make the region competitive in an increasingly globalising world.

A Model of Knowledge Management for Tourism Destinations

If we accept this book's notion of destinations as networks of stakeholders, then the question for knowledge management is to how to improve the knowledge base of the destination to deliver competitiveness. In order to effectively define, develop and diffuse knowledge through a network of heterogeneous stakeholders in a destination, a sound conceptual framework is essential. Here, any model of knowledge management must ensure that the process aligns with, and contributes to, the goals of the organisation or destination. This allows knowledge management projects to match destination objectives. It also demands a clear identification of what knowledge is important to support the long and short-term needs of the destination. Of course, there are many models of knowledge management that could be useful to tourism. However, the root of all of these models is a structured knowledge stocks and transfer approach. Here, *knowledge stocks* are the things that are known and *knowledge transfer* is the means by which the knowledge is communicated to those who need it, and so how learning across the destination is achieved. This model works well for tourism and offers insights and practical techniques to facilitate the creation and transfer of tourism knowledge from researchers to the sector.

Knowledge transfer

For the contemporary tourism destination, knowledge transfer is essential for competitiveness and there is an increasing trend for destinations to intervene in the process rather than to remain passive (OECD, 2001). Yet, the effective transfer and use of knowledge is not an easy task – indeed, estimates suggest that whilst tacit knowledge accounts for 80% or more of the knowledge held in an organisation, only 10 or 20% of that knowledge is utilised in a transfer process. Here, the knowledge management literature provides insights into understanding the knowledge transfer process. Of course, knowledge transfer at the destination can happen informally through spontaneous or unstructured processes, but a well-managed destination does not leave it to chance, and manages and thinks through the process. Here, the key element is the imperative of transfer plus absorption, and a range of tools and activities can be employed to make it happen. Transfer occurs through peer-to-peer exchanges, iterative knowledge sharing, team learning and electronic discussion spaces. The core concept is to ensure the effective application of intellectual capital in a tourism organisation or destination to achieve objectives and, over time, there has been a shift from creating knowledge repositories to integrating knowledge sharing and transfer into core business processes. Nonetheless, for effective transfer to occur, organisations must participate and be embedded within destination networks, as partners can control degrees of access to knowledge.

Knowledge transfer within destination networks

Whilst it is recognised that knowledge transfer across tourism destinations is important, compared to other fields the tourism sector is relatively undeveloped. As a result, tourism has not, until recently, been subject to a knowledge management approach and destinations are not as competitive as they could be. This, then, is the challenge for tourism. The knowledge-based economy is not simply the preserve of hi-tech industries, and in fact represents the use of knowledge to benefit all sectors of the economy, including tourism.

There are two reasons for this failure. Firstly, tourism knowledge generators – researchers, consultants and others – belong to a 'community of practice', with common publications and language – a community of practice that does not include the practitioners at a destination. This concept of different 'communities of practice' may be partly to blame for the lack of gearing between those that generate knowledge and practitioners at tourism destinations. Whilst this goes some way to explaining the poor record of knowledge transfer in tourism, other authors state that in addition, many of the prior conditions necessary for the successful transfer and adoption of knowledge are not present in tourism. This problem is related to the very nature of tourism destinations. Destinations are characterised by:

- A dominance of small enterprises which are often single person or family owned. As a consequence, knowledge must be highly relevant to their operation if they are to adopt and use it.
- Fragmentation across a variety of sectors – accommodation, transport, food and beverage.
- Vocational reinforcers, such as poor human resource practices, which militate against the continuity of knowledge absorption. These include the employment of seasonal and part-time workers, high labour

turnover and a poorly qualified sector which inhibits knowledge absorption. It is therefore more difficult to achieve effective knowledge transfer to employees who have a low commitment.

If knowledge transfer is to be effective in tourism destinations then the process needs to be mapped and understood. Hjalager's (2002) model of four channels works well, with knowledge transferred according to the sector of tourism and the use of the knowledge. Her four channels are:

(1) The technological system.
(2) The trade system, where transfer takes place through trade associations and tends to be sector or destination-based.
(3) The regulatory system, where knowledge of, say, fire regulations is transferred.
(4) The infrastructure system, including managers of parks and natural resources, where there is a greater tendency to accept and use knowledge.

The knowledge transfer approach can be likened to the concept of diffusion of innovations across a destination, where knowledge is viewed as an innovation. Rogers and Shoemaker define an innovation as anything:

> perceived as new by an individual, and it matters little…whether or not an idea is relatively new … it is the perceived newness of the idea for the individual that determines his reaction to it. If the idea is new to the individual, it is an innovation. (Rogers and Shoemaker, 1971: 19)

Rogers (1995) defines diffusion as the process by which an innovation or new idea is communicated through certain channels over time among the members of a social system. Effectively, the decision to adopt an innovation is the outcome of a learning or communication process; knowledge is the 'dynamic' in the system and it also binds the elements of the innovation together. This approach has much to offer the knowledge transfer process at the destination, particularly in terms of understanding the success factors of diffusing an innovation and securing its adoption.

For knowledge transfer a fundamental step is the identification of factors related to the effective flow of knowledge and of the characteristics of knowledge flows, knowledge reception and resistances to adoption. The diffusion literature identifies the following elements of the system as critical in the process:

- The sources and legitimacy of knowledge, as well as the quality and reliability of the knowledge.

- Adopter characteristics and capacity to adopt. In other words, the point to which knowledge is transferred to, and how it is deployed. In addition, other relevant factors include organisational size, structure and competence of the adopter.
- Different problem domains (such as routine/non-routine or complex/ basic) will demand different transfer techniques.
- The nature of the media used for the transfer must be appropriate to the innovation – in particular the use of 'relational' channels, where it is possible to develop the frequency and depth of two-way human-to-human contact.
- The degree of partner similarity in terms of interests, background, or education. This relates to the idea of 'communities of practice' mentioned above and works at both the individual and the organisational level.
- The level of depreciation of knowledge after transfer.
- The level of organisational self-knowledge – the more an organisation knows, the more receptive it tends to be.

The knowledge transfer approach

The key to understanding the transfer of knowledge across a destination is to recognise the importance of the type of knowledge to be transferred and the medium to be used for the transfer. It is therefore important to accurately assess the type of knowledge to be transferred and to understand the destination context within which it is exchanged. Chua (2001) examined the transfer of three types of knowledge. His three types of knowledge were as follows:

(1) *Codifiable*, or effectively explicit knowledge, as measured by the extent to which the knowledge can be articulated or represented in documents and words. This knowledge may be substantive, for example, in blueprints, or it may be procedural, for example, in a recipe for carrying out a task. The more explicit the knowledge is, the greater is its ability to be codified.
(2) *Teachable* – the more tacit the knowledge, the harder it is to teach. Teachability is the ease by which the knowledge can be taught to another person.
(3) *Complexity* – the more elements needed to complete a task, the more complex the knowledge. Complexity refers to the number of critical and interacting elements of the knowledge needed to accomplish a given task. The more elements needed to complete a task, the greater is the complexity of the knowledge.

Chua categorised media channels used for transfer by their degree of *richness*, where:

> the media richness of a channel can be examined by its capacity for immediate feedback, its ability to support natural language, the number of cues it provides and the extent to which the channel creates social presence for the receiver. (Chua, 2001: 2)

These dimensions are explained by Chua (2001) as:

- *Immediate feedback*. This refers to the channel's capacity to deliver a certain amount and promptness of feedback to the sender. For example, face-to-face discussion is highly interactive.
- *Natural language*. A channel is regarded as having the ability to support natural language if the sender can structure and send the message in the most intuitive manner or as if it were in a conversation.
- *Cues*. The number of cues or senses provided by the channel includes both verbal and non-verbal cues such as tone of voice, hesitation, facial expressions, vocal cues, dress and posture. These cues help the individuals to interact more effectively.
- *Social presence*. The extent to which the channel can be used to create social presence is closely related to the number of cues provided. When a message receiver feels that the sender, rather than the medium, is actually delivering the message, the channel is said to afford a high social presence. The social presence provided by a channel influences individuals' motivation to engage in interpersonal communication.

Chua found that the more explicit the knowledge, the less rich are the media used to transfer it. Conversely, the richer the knowledge, the more technology is needed in the transfer process. Other research has also examined the media used for knowledge transfer and sharing. For example, Lionberger and Gwin (1991) found that rapid increases in adoption rates occur mostly through the medium of people talking to and influencing each other. Similarly, Johnson (1996) states that for knowledge sharing, face-to-face interpersonal communication is the preferred mode of communication, particularly for information-seeking. In a study of dissemination amongst small and medium sized enterprises (SMEs), it was found that the medium of peer networks is more valuable than consultants and other change agents, as SMEs prefer to have contact with other people who are doing the same thing, of whom they can ask questions freely (Friedman & Miles, 2002). Therefore Martin (2002) finds that SMEs can benefit from entering into alliances as a means to achieving mutually beneficial objectives and establishing peer networks.

Of course technology is at the heart of effective knowledge transfer and sharing. As we move into an era of e-knowledge, technology facilitates more rapid transfer. In particular, the World Wide Web is the perfect medium for the easy mass distribution of information across a destination network, as well as the use of intranets. This technology is the foundation of destination-wide systems of knowledge transfer and sharing, not only within the destination, but also with customers and suppliers worldwide.

The most effective use of technology for knowledge transfer and sharing now lies in the use of knowledge portals – an approach that is being adopted by many destinations. Portals are a powerful medium for transferring knowledge. They provide an integrated framework linking users with knowledge in a single point of access, effectively providing a 'virtual workplace'. Portals use content management tools, online collaboration tools, are secure and can be both customised and personalised. Essentially, portals are becoming the centrepiece of knowledge management systems for destinations, linking the technology and the people elements and allowing the producers and users of knowledge to interact.

Knowledge transfer models for destination networks

There are three models of knowledge transfer that are useful to destination networks:

(1) At the level of the destination organisation, the transfer of knowledge within *knowledge creating organisations* refers to the process where tacit knowledge is transformed into explicit knowledge through the interaction of employees and management in cross-functional teams. The focus here is on interaction between individual knowledge and organisational knowledge. This approach has been championed by Nonaka (1991).

(2) At the destination level, *organisational learning* refers to a method where organisations build competitive advantage through effective transfer management and constant updating of information. It requires the development of knowledge transfer structures that stress learning agents within the organisation who respond to and communicate internal and external information to co-workers. In tourism this approach includes outreach strategies and networking concepts within communities of practice. This relates to one of the oldest forms of transfer used in tourism – that of 'extension'.

(3) For tourism destinations, the model of *absorptive capability* is very relevant, as many of the users of tourism knowledge lack experience in the field. SMEs, for example, will only use knowledge if it is highly relevant to their operation. Here, the idea of organisational capability is about filling the 'gap' between intention and outcome by making tacit knowledge explicit. The absorptive capability model refers to the fact that organisations have to respond to inputs of knowledge and that their ability to do so will depend in part on the organisation's existing knowledge; effectively, the greater the knowledge stocks, the more effective will be the assimilation of new knowledge. It will also depend upon the size, internal structure, division of labour, leadership and competency profile of the receiving organisation. Clearly then, destinations will vary in their ability to engage in knowledge transfer and this will in turn impact upon their performance and competitiveness.

From the user's point of view, this model considers the capacity to acquire and apply knowledge, the ability to build or improve upon knowledge and to transform as much knowledge as possible for use in decision-taking.

The key to the model of absorptive capacity is the nature and characteristics of the organisation receiving the knowledge. Ladd and Ward (2002) have put forward a rationale to explain which organisational cultures might be more or less conducive to knowledge transfer. To quote Ladd and Ward (2002: 3) these cultures can be characterised as follows:

- *Openness to change/innovation*. This type of organisation has a culture characterised by openness to change and innovation. The organisation is more likely to encourage human-to-human contact and knowledge transfer and to develop the talents and individual knowledge of its workers. This type of organisation would be expected to be an excellent environment for knowledge transfer and for tourism.
- *Task-oriented*. This type of organisation has a culture that is similar to the 'openness to change/innovation', but it fosters a shared philosophy which encourages the convergence of the goals shared by the organisation and its employees. This type of organisation stresses quality and has an attention to detail that tries to enhance the efficiency of knowledge transfer. This type of organisation would also be expected to be a positive environment for knowledge transfer. In tourism, this type of organisation is often found in the public sector, the education sector and in the hotel sector.
- *Bureaucratic*. This type of organisation discourages interpersonal communication and so will diminish the relational channels that are

so effective in knowledge transfer. In addition, this type of organisational culture is likely to discourage the pursuit of individual knowledge and so will not be a positive environment for knowledge transfer. In tourism, the more traditional public sector-based destination management organisations can fall into this category.

- *Competition/Confrontation.* This type of organisational culture tends to discourage interpersonal relationships. It is also a culture that fosters a pursuit of power and so may put individual goals at odds with organisational goals. It is therefore likely to create a negative environment for knowledge transfer.

Applying Knowledge Management in Destination Networks

There are to date only a small number of examples and applications of knowledge management across destination networks (Richards & Carson, 2006). However, recognition of the significance of the approach is growing as practitioners recognise the value of knowledge sharing not just within organisations but also across destination networks, and in particular the encouragement of partnerships within the destination (see for example Fesenmaier & Parks, (1998); Gretzel & Fesenmaier (2002); Micela *et al.* (2002); Pechlaner *et al.* (2002) and Peters *et al.* (2002). It is characterised by the fact that the early phases of knowledge management were dominated by the phrase 'knowledge is power'. The new thinking argues 'sharing is power', and this creates 'communities of knowledge' at the destination level. However, it must be recognised that, at the destination level, learning between organisations needs certain conditions to be effective. These conditions include:

- Continuous innovation for competitiveness and new destination products.
- Interactive knowledge sharing, facilitated by the Internet and 'destination knowledge portals'.
- Stable relationships with high levels of trust. For tourism SMEs this may require a period of training and induction.
- Effective knowledge processes within organisations to generate knowledge stocks.

This new thinking has given rise to the concept of sharing 'knowledge capital' and relates back to discussions on the nature of knowledge as a 'resource'. Knowledge itself does not deliver growth; it has to be incorporated into the production of goods and services because it is linked to human capital and labour markets. Here, networks of organisations at the destination recognise that the same knowledge can be used by more than one organisation but that the cost does not increase in terms of accessing that knowledge. This is reflected in the fact that tourism enterprises are showing increased interest in cooperation across organisational and geographical boundaries. In other words, the traditional notion of the enterprise as a closed, self-regulating system is being complemented by a broader conceptualisation that includes ideas such as strategic alliances, outsourcing, virtual enterprising, and externalisation of resources. Effectively, the enterprise has become a networked organisation and its success will depend upon the degree to which it can leverage strategic advantage from its networks. For tourism destinations, this concept is best illustrated by the dominant form of organisation in tourism – small/medium sized enterprises (SMEs).

Knowledge management in destination networks of small businesses

The dominance of SMEs in the tourism sector provides a challenge for the implementation of knowledge management at the destination (Braun & Hollick, 2006); indeed, it could be argued that tourism, and in particular small tourism businesses, are hostile environments for knowledge management. Here there are three elements that need to be considered:

(1) Understanding the adoption of knowledge management in destination networks of tourism SMEs.
(2) Developing strategies to increase the acceptance of knowledge transfer and sharing.
(3) Utilising stakeholder and social network techniques to better understand the flows and adoption of knowledge in the destination.

Magnusson and Nilsson (2003) examined knowledge management approaches in networks of SMEs. They observed two contrasting types of network:

(1) *The supply-chain network.* This network displays a low degree of knowledge integration for a number of reasons. Firstly, the network is usually created from a previous collaboration along a supply chain that has evolved towards a network and where all partners have an understanding of what the end-product is, based upon self interest. Secondly, there is an overtly instrumental relationship between the partners in the network, with the primary communication and knowledge exchange directly tied to production. This leads to a low integration of knowledge, as the enterprises see no need for knowledge exchange in areas that are not directly relevant to the functioning of

the supply chain. For destinations, the knowledge coordination in these types of 'instrumental networks' tends to be done by industry associations or by the larger enterprises at the destination.

(2) *The research network.* This network contrasts with the supply-chain network as it displays a high degree of knowledge integration between the networking partners. This is because the network is characterised by knowledge creation as the primary objective for the collaboration. The network usually comprises a collection of enterprises with certain expertise that collaborate in some type of research activity, but where the outcome of the process is not easily defined. For destinations, knowledge management coordination tends to be public sector-led (often by the destination management organisation or parent body) (Richards & Carson, 2006). They provide an arena for the management of knowledge, such as the acquisition and implementation of information and communication technology and other structural investments, as well as the creation of knowledge-sharing cultures.

Given these two generic types of SME networks, Shapira and Rosenfield (1996) recommend a series of strategies to promote and accelerate knowledge transfer and sharing amongst SMEs:

- Awareness building and demonstration to make potential users more knowledgeable about available knowledge sources, their possible applications, and their benefits and costs.
- Information search and referral services to encourage knowledge sharing and reduce information search costs.
- The use of experts to assess business problems, identify opportunities to use and share knowledge and assist in implementation.
- SME training programmes, which may be conducted through on-the-job training, classroom training, management seminars, team-building workshops, and distance learning. This is particularly important in building the trusting and cooperative environment for knowledge transfer (Florida, 1995).
- Personnel exchange and the support of knowledge workers, as the SMEs themselves may lack the internal capability to absorb knowledge management approaches and resources.
- Inter-firm cooperation to resolve common problems and share information and learning, achieve scale economies in service provision and technology deployment, and strengthen ongoing business and technology development.

- Regional or sectoral cluster measures can be used as an opportunity to strengthen organisational capabilities and linkages within particular regions and industrial sectors. Levels of communication and dialogue between technology developers and users and among users, institutional credibility and leadership, and other aspects of social capital have been shown to be extremely important in the diffusion of technology.

Policy Implications for Destinations

There is a view that informal governance structures develop within networks because the policy environment has lagged. These loose governance systems at the destination level therefore act as an alterative to the public sector. For tourism destinations, an important notion is that of networks having their own embedded macro-culture, with the behaviour of stakeholders controlling the degree of access to knowledge of the various participants (Pavlovich, 2001). To counter these informal governance mechanisms, governments have been faced with the need to develop policy initiatives, including those at the destination level, in order to effectively deal with the process of knowledge creation and transfer. Effectively, these policies grapple with the issues surrounding the nature of knowledge as a public good. These issues include access to knowledge, the removal of barriers to knowledge transfer and adoption and the need to encourage private enterprise to share knowledge. In many cases, policies adopt the network concept by intervening at the level of knowledge creating or receiving node (the organisation), or facilitating knowledge flow and transfer around the network. The OECD for example recommends:

> Policies that engage human capital, innovation and entrepreneurship in the growth process alongside polices to mobilise labour and increase investment, are likely to bear the most fruit over the long term. OECD (2001: 8)

This is an important new policy context for tourism destinations and is discussed in detail in Hjalger (2002: 473). She states that it is an issue that must be addressed if tourism is to be 'a professional and respected stakeholder in economic life'. However, it must also be said that much of this debate and the utilisation of knowledge management has passed tourism by, and it is only in the 21st century that the tourism sector and researchers have begun to realise the significant potential of the approach. To date there is little that focuses upon the policy implications of knowledge flows and networks at the destination level. Instead, most of the literature is focused on the more generic 'regional' scale, although this still has relevance to tourism. For example, the OECD (2001) has published an influential report on knowledge policy at the regional level emphasising a shift from the national to the sub-national level in terms of innovation and knowledge policy. This shift is due to the growth of urban and regional development authorities as well private/public sector partnerships. These agencies commonly have tourism powers. This means that new forms of governance and policy for the knowledge economy are needed at the destination level. The real tension here is between global and local influences, and the competitive advantage for destinations lies in ensuring that regional differentiation is protected. At the regional level, policy intervention for knowledge development and transfer is needed to ensure:

- Learning takes place at the individual and the organisational level.
- The interaction between organisations takes place at the destination level in terms of knowledge sharing, otherwise they may have no incentive to do so. The most supportive environment for this is one of collaboration and consensus, with public/private partnerships creating 'networked destination governance'.
- An equitable distribution of benefits around the destination, including the host community.

The production and distribution of knowledge are increasingly significant processes determining economic performance and competitiveness at the regional and destination level. However, it might be expected that regionally, their impact will be differential. This may be due to the fact that interaction between organisations varies according to the social and spatial configuration of each region. This is clearly illustrated by the high level of innovation in California's Silicon Valley, where the spatial proximity of universities and companies sparked innovation. However, given the fact that the success and impact of policy may vary according to the region itself, it begs the question as to whether the region (or destination) is the best level to apply knowledge and innovation policy. In fact, to be successful there needs to be a careful application of general policy principles to particular economic and social regional circumstances at the destination level.

Conclusion

This book is based upon the view of destinations as large complex network systems with many levels of interaction between stakeholders. This contemporary destination is undergoing processes of fundamental change. These include shifting governance structures, the impact of technology and the belated recognition of the importance of knowledge and innovation to destination competitiveness. For destinations to be successful and competitive in the future, they will need to innovate and develop new products. The basis for such innovation is the generation and use of new tourism knowledge. By combining the concepts of the network approach to destinations with the insights provided by knowledge management it becomes clear that a new policy environment will be required to ensure the sharing and adoption of tourism knowledge across destination networks. These policies will encourage the use of effective models of knowledge creation and transfer allowing destinations to respond flexibly and quickly to threats as they develop destination knowledge stocks. In other words, the 'learning destination' becomes a reality, characterised by processes of mutually reinforcing interaction and cooperation between stakeholders delivering a collaborative and competitive destination.

This book shows that not only can network analysis help to understand the structuring of destinations, but also this chapter provides insights into developing effective strategies for knowledge sharing and adoption, in particular by identifying barriers to knowledge adoption such as weak or broken links. Additionally, by analysing the key stakeholders in a destination involved in the production of knowledge stocks in different problem domains (such as say, planning or marketing), it is possible to understand the key knowledge gaps at the destination. It is essential that these processes are understood if destinations are to remain competitive, innovative and develop new products. Research in knowledge management and destination networks and stakeholders is potentially both a rich and valuable area for tourism researchers. Not only is it an area that examines phenomena of competition and cooperation that are commonly found in tourism destinations, but it also offers real promise for improving destination competitiveness.

Chapter 6

Policy Networks and Tourism Governance

DIANNE DREDGE and CHRISTOF PFORR

Introduction

Since the 1980s, structures and processes of government have been transformed, leading to new forms of networked policy-making and governance. In many western democracies, inquiries into public administration led to increasing criticisms of traditional, bureaucratic models of governing (e.g. Commonwealth Government, 1983; Wilenski 1977, 1982). Critics argued that rational scientific approaches to public policy-making were inefficient and centralised bureaucracies stymied change (Pressman & Wildavsky, 1984). In a rapidly globalising world economy, governments were diagnosed as slow in addressing many complex policy issues and ineffective in balancing broad-ranging corporate and public interests (Amin & Thrift, 1994). These criticisms stimulated renewed interest in democracy and public participation (e.g. Giddens, 1998; Ladeur, 2004) and a reinvention of government structures and political governance (e.g. Howlett & Ramesh, 1995; Rhodes, 1997; Klijn & Koppenjan, 2000). In tourism, as with many areas of policy, old forms of centralised, bureaucratic policy-making were replaced with new forms of interactive governance, collaboration and partnerships (e.g. Bramwell & Lane, 2000; Hall, 2000; Pforr, 2005). Yet despite the rise in networked tourism policy-making, questions about the utility and implications of policy networks remain. Are these new networked approaches more efficient and effective in producing tourism public policy? Do networks promote better tourism governance? What are the advantages and disadvantages of tourism policy networks?

This chapter explores these questions by interrogating the nature and characteristics of policy networks and their application in tourism. Our objectives in this endeavour are twofold. Firstly, we seek to progress an understanding of the significance and implications of tourism policy networks by drawing together the wider landscape of network theory and existing case study research in tourism. Secondly, we aim to highlight the strengths and weakness of existing research and provide directions for tourism policy network management. The advantages and disadvantages of networks as a policy-making approach will be discussed in terms of recent emphasis on democratic, transparent and accountable government practice and the third way politics project.

What are Policy Networks?

Policy networks are sets of formal and informal social relationships that shape collaborative action between government, industry and civil society (e.g. Atkinson & Coleman, 1992; Howlett & Ramesh, 1995; Rhodes, 1997). They are characterised by a variety of participants whose actions and interests transcend organisational boundaries (e.g. Howlett & Ramesh, 1995; Rhodes, 1997; Marsh, 1998). Policy networks involve cooperative commitment by network members to a set of common goals, sometimes beyond their immediate self-interest goals, in order to help chart the position and activities of government (Burstein, 1991). Moreover, Rhodes (1990) observes that over time, stable policy networks can develop quasi-institutional structures and rules of conduct, and can become directly enmeshed in government policy-making and implementation such that lines of authority and responsibility are blurred. The rise of networked policy-making has been mooted as an opportunity to establish more collaborative, democratic and inclusive policy-making. It is this question of the role and influence of policy networks in policy-making that is the focus of this chapter.

The Rise of Policy Networks

The shifting nature of government policy-making provides the context in which to understand the growth of tourism policy networks and their potential impact on policy-making (Held, 1989; King, 1990; Rhodes, 1997; Pierre, 2000). Public policy-making in the early 20th century was driven by an expansion of state responsibilities and, by corollary, growth in bureaucratic policy-making. The ideas of Max Weber (1922) and John Maynard Keynes (1936) were instrumental in the growth of government over this period. While we are keen to avoid over-simplification of the enormous influence that Weber and Keynes had on thinking about government and its role in economic management, Keynes' idea that government had an important role intervening in economic affairs to stimulate employment and economic growth contributed greatly to the expansion of government activity into diverse policy arenas. Similarly, Max Weber's ideas about the separation of political and bureaucratic arms of government and the contribution that a professional public service could make to the rigour and rationality of government policy-making also influenced the expansion of bureaucratic structures and processes. As a result, over the early to middle part of the 20th century, bureaucracy grew in terms of its size and its policy reach as new policy issues emerged. Tourism, the environment, transport and local economic development are examples of policy areas where government became increasingly active.

As pressure to secure economic growth continued unabated over the course of the 20th century, dominant ideologies about the nature of policy-making began to shift. Criticisms of heavy-handed government intervention and associated market inefficiencies, described as the 'Keynesian legacy', led to renewed interest in the liberal ideologies originally articulated by Adam Smith in the 18th century (Dredge & Jenkins, 2007). Whilst this neo-liberalism diverges from its earlier roots in various ways, the driving belief is that non-interventionist government and free markets can supposedly deliver better, more efficient outcomes for the public good. Following this line of reasoning, governments adopting neo-liberal management philosophies have sought to reduce their direct involvement in many areas once considered to be the domain of government. It is more effective, they argue, to facilitate private sector investment and enable communities to take responsibility for their own needs. As a result, governments have downsized, shifted responsibilities to other levels of government, and opened their activities up to market competition (Davis & Weller, 2000; Hughes, 2003). This withdrawal of direct government involvement in many areas of policy has, in turn, opened up opportunities for non-government sectors to undertake many traditional roles and activities of government. But with these changes have come questions about what interests are really embedded in this new way of policy-making.

Amin and Thrift (1994) argue that, from the 1980s onwards, these changing roles of government have led to economic restructuring on a global scale. There has been an increasing flightiness of capital across international borders and, as a result, localities are more vulnerable to conditions outside the control of governments. Issues once considered public are now characterised by complex webs of relations between government and non-government interests, and governments must now work collaboratively with non-state actors to manage complex public–private sector issues (Schneider, 2005: 7). Government involvement in tourism has not escaped these trends, illustrated in the corporatisation of state tourism departments in many countries in the 1970s and 1980s (e.g. Jenkins, 2000; Hall, 1999). In tourism it was considered that the private sector was best suited to managing tourism and, as a result, corporate entities and statutory corporations significantly or totally funded by the public purse were established. Moreover, industry representatives were appointed to governing boards and they were managed on a commercial basis, thereby raising questions over what interests were being served (Jenkins, 2001).

This shift from bureaucratic models of policy formulation to those that involve downsizing, commercialisation, outsourcing and consultation with multiple stakeholders is referred to as the shift from public administration to public management (e.g. King, 1990; Hughes, 2003). These new models of public management rely less on formal political–administrative structures and rational scientific models of policy-making and more on networks of collaboration and public–private partnership formation. Ladeur (2004) observes that this change also involves a shift from objective rationality to one of relational rationality – i.e. where decisions are based on the subjective, constructed rationalities of participants and not upon some bureaucratic construction of what is in the public interest. In the above context networked policy-making has emerged as a new form of 'decentred governance based on interdependence, negotiation and trust' between government and non-government interests (Sørensen & Torfing, 2005: 196).

Historical Development of the Policy Network †Theory

Clusters of interest in policy-making were observed in early 20th century political science, but the theoretical development first emerged in Heclo's (1978) observations about the locus of policy-making power in the US political system. Heclo observed that power was not vested solely in the executive, but that there were 'iron triangles' of influence that incorporated small numbers of actors in stable relationships. Later Sabatier (1987) and his colleagues argued that there were multiple, dynamic constellations of interests and that advocacy coalitions formed over certain issues to influence public policy and government action. Sabatier's work complemented European research emerging during the early 1990s that focused on the structure and function of networks (Atkinson & Coleman, 1992; van Waarden, 1992; Schubert, 1991). Pappi (1993) characterises the developments since the early 1990s as a 'boom' for policy networks, while König (1998: 387) observes 'policy network analysis has become a dominant paradigm for the study of public policy'. This remark is, however, Eurocentric in origin, and does not acknowledge the patchiness of attention to networks in other parts of the world.

Policy network theory is a response to intense debate within policy studies in the 1970s and 1980s. Traditional approaches to policy-making were accused of portraying policy-making as far too schematic, too bureaucratic and too removed from the irretractable influences of power, values and interests and institutional cultures (see Heclo, 1978; Sabatier, 1987; and Wright, 1988 for early discussions that provide the basis for the development of network theory). Policy analysts began to reject the notion that policy-making was a rational scientific process comprised of neat, sequential stages (Hogwood & Gunn, 1984). As this messy reality of policy development has come to dominate in recent years, policy network theory has contributed to explanations of dynamic, complex and unordered policy processes (Héritier, 1993; Pforr, 2005; Schneider, 2005). In this context, Atkinson and Coleman point to the usefulness of networks as a theoretical concept because it is:

> ...both encompassing and discriminating in describing the policy process: encompassing because they refer to actors and relationships in the policy process that take us beyond political-bureaucratic relationships; discriminating because they suggest the presence of many communities and different types of networks. Atkinson and Coleman (1992: 154)

Accordingly, policy networks have emerged as an alternative explanation of public policy-making, which complements or even replaces traditional bureaucratic models. Policy network theory reinforces the idea that policy formulation and implementation comprises networks of interdependent actors, committed to a particular set of ideas or objectives, and continuously engaging in the identification, framing, discussion and negotiation of policy issues, problems and opportunities. As Sørensen and Torfing highlight, these networks:

> ...can either be self-grown or initiated from above. They might be dominated by loose and informal contacts or take the form of tight formalised networks. They can be intra- or inter-organisational, short-lived or permanent, and have a sector-specific or society-wide scope. The multiple forms of governance networks attest to the broad relevance of the concept for describing contemporary forms of societal governance. (Sørensen and Torfing, 2005: 197)

The attractiveness of the network approach in understanding policy-making is that it draws from pluralist and corporatist theories of policy-making, in that multiple interests are incorporated into policy formulation and implementation (Rhodes, 1997; Howlett & Ramesh, 1995; Thompson & Pforr, 2005). But the network approach also differs from these theories in fundamental ways. Most notably, pluralist and corporatist approaches conceive government as the central actor that prejudices some interests – corporate or plural – over others. In the network approach, government does not have an overarching instrumental decision-making role. Instead, decision-making is shared and the power to make a difference is distributed amongst relatively autonomous, non-hierarchical actors and agencies with an interest in the policy issue (Atkinson & Coleman, 1992; Schumann, 1993; Mayntz, 1993; Héritier, 1993; Rhodes, 1997; Thatcher, 1998).

In the context of tourism, network theory provides two important streams of understanding for the study of tourism public policy development and for understanding more about government–industry–community relations (e.g. Tyler & Dinan, 2001; Pforr, 2002, 2006a; Thompson & Pforr, 2005; Dredge, 2006b). These streams include:

(1) *Literature that seeks to describe and analyse organisational structures and relational characteristics of networks.* This body of work examines the structure and function of networks and is characterised by a variety of criteria, typologies, models and frameworks for describing and analysing policy networks.
(2) *Literature that examines the roles and management of networks.* This stream of literature is heavily influenced by growing discourses on (a) participatory democracy and governance, and (b) business planning and management.

The former examines the characteristics, role and implications of networks on government–business–civil society relations, the political nature of planning and policy-making and issues of network management. The latter explores network management associated with maximising competitive advantage, commercial complementarity and tourism product packaging.

Dredge (2006a) observes that there is growing attention to the descriptions and analysis of organisational structures and relational characteristics of networks where such applications provide important insights into how business relationships are formed and managed, and how clustering and complementarity within the industry can be maximised. There has not been any consolidated attempt by tourism researchers to explore the implications of networks as a form of governance and the effect that networks have on tourism policy-making practice. Most research has tended to be fragmented case study research and greater attention needs to be placed on wider observations and theoretically informed conclusions. This is an important aspect that will be returned to later in this chapter, but before doing so it is important to briefly discuss some of the advantages and disadvantages of network theory that have emerged in the wider literature.

Advantages and Disadvantages of Policy Networks

A number of important disadvantages and advantages have emerged that need to be addressed if networks are to add value to policy-making practices (e.g. see Dowding, 1995, 2001; Börzel, 1998; Klijn, 1996; Klijn & Koppenjan, 2000; Marsh, 1998). Disadvantages and criticisms include:

(1) *Definition and clarity of theoretical concepts is lacking.* Policy network research has been conducted principally in the US and Europe and divergent definitions of fundamental concepts (e.g. policy networks, policy communities and governance networks) have thwarted an integrated approach to understandings (see Wright, 1988; Börzel, 1998).

(2) *Policy network theory is predominantly descriptive and lacks explanatory power.* Network theory is useful in describing the characteristics of networks but does not provide insights into cause–effect relationships, nor do they explain the outcome of policy processes (e.g. Dowding, 1995; Blom-Hansen, 1997; Börzel, 1998).

(3) *Issues of power, conflict and representation are neglected in network theory.* Networks focus on examining the nature of communication and collaboration and offer little insight into the role of conflict and power inequalities (see also Bramwell, 2004; Thompson & Pforr, 2005; Dredge, 2006b). Networks allow particular interests to be prejudiced over others in policy formulation. The opportunity to give weight to some interests over others also allows policy innovation and entrepreneurialism to be stifled.

(4) *Network theory neglects methods and criteria for evaluation.* Network theory is based on the social constructionist view of policy-making where policy emerges from social relations and dialogue between actors. As a result of this emergent nature of public policy, network theories do not provide clear frameworks for evaluation (e.g. Klijn & Koppenjan, 2000; Thompson & Pforr, 2005).

(5) *Network theory lacks a normative dimension in that it does not inform network formation or management.* As a result, it is unclear how public interest is treated and, as guardian of that interest, what government role is (Mayntz, 1999; Sørensen & Torfing, 2004).

But these criticisms must be balanced against a number of advantages identified and discussed by Dredge (2006a). Firstly, policy networks recognise the overlapping and simultaneous manner in which different issues within the one policy community can be addressed by different networks operating at different scales and over time. For example, networks addressing regional coordination, the development and management of tourism product, and marketing and promotion may all co-exist within the one tourism policy community. Secondly, the network approach recognises that the distinction between private and public is blurred. The network approach therefore fits well with the realities of tourism as a multi-dimensional area of public and private sector policy interest. Thirdly, the network approach recognises that different levels of political support may exist for different policy issues within the one policy network. For example, there may be political support for the development of a local tourism association but the same level of support may not exist for regional cooperation. Fourthly, the network approach recognises that policy actors may have membership in different policy networks and their powers, roles, functions and level of support and interaction may vary within these structures.

These advantages and disadvantages require critical attention if network theory is to move beyond its current limitations as a descriptive and analytical toolbox and metaphor for complex socially constructed policy-making. In an effort to explore how the power of network theory can be harnessed to provide better policy and planning outcomes, an examination of the role of government, its relationship to and involvement in networks, conceptions of public interest and democracy has been suggested as a way forward (Klijn & Koppenjan, 2000; Sørensen & Torfing, 2005).

Moreover, these advantages and disadvantages suggest that a number of important questions about policy networks and their contribution to 'good' policy-making practice remain unanswered, including whether policy networks facilitate 'good' policy output and outcome and democratic political governance (see Klijn & Koppenjan, 2000). This brings into focus questions about the relationships between government, networks and public interest previously identified.

Government, Networks and Public Interest

How public interest is defined, and the nature of relationships between the state, business and civil society, is being increasingly scrutinised in contemporary policy studies. The growing diversity of stakeholders, the rise of individualism, the emphasis on corporatist agendas, the uptake of neo-liberal economic management philosophies by some governments, the waning confidence in government, and the increasing complexity and interrelatedness of policy problems have forced governments to reassess the way they define, value and address public interests (Davis & Weller, 2000; Hughes, 2003). In a powerful exposé of the renewal of social democracy, Giddens (1998: 64) argues for a 'third way politics' '…to help citizens pilot their way through the major revolutions of our time: globalization, transformations in personal life and our relationship to nature'.

Giddens (1998: 71) argues that the crisis in democracy is that it is not democratic enough, identifying a number of values on which the third way politics should be based. Firstly, he argues, third way politics should be concerned with social justice, equality and freedom. He acknowledges that collectivism and overarching definitions of public interest are problematic, and that freedom to pursue individual dreams and take control of one's life is paramount to societal improvement. New relationships between individuals, communities and governments are needed. These new relationships should embody the rights of individuals and communities to have a role in decision-making, but that there should be 'no rights without responsibilities' (p. 65). That is, individuals and communities must be responsible in the way that they engage in civil society and how they pursue their interests.

The second precept on which Giddens bases his 'third way politics' is that there should be 'no authority without democracy' (p. 66). The renewed emphasis on social justice, equality and democracy transfers authority to individuals and communities to make decisions, but this authority should not be exercised without full and active participation of the broadest sections of civil society. The re-democratisation project acknowledges the contribution of small groups and voluntary associations in getting things done, but that civic engagement has traditionally involved 'the more affluent strata' of civil society (p. 84). To counter the emphasis on these strata, Giddens calls for governments to repair the imbalance, to address the failings of neo-liberalism, and to encourage bottom-up decision-making and new forms of local autonomy.

In this third way politics project, Giddens opens up important criticisms of contemporary relationships between civil society and the state and puts forward a project to reshape democratic participation. This reshaping advocates a middle ground between neo-liberalism and the socialist left which is appealing but vague and oft-criticised for its lack of substantial direction (e.g. Jary, 2002). Increasing attention on governance structures and practices that recognise political life, solidarity, shared goals, dialogic democracy and the role of the state go hand in hand with this third way politics.

Governance refers to the relationships between the state, civil society and economic interests through which decisions are made that 'steer' a society and produce social order (see Börzel & Risse, 2005). It involves multiple interest groups engaging with a more open and transparent government than has traditionally been the case in centralised bureaucracies. In an ideal situation, policy networks as a new form of governance will be democratic if all members of the network are afforded equal opportunities to participate in and influence political decisions. Of course, this occurs rarely, if at all, so there is a need to investigate and understand the influence of networks on the democratic nature of policy-making and their impact on legitimacy and transparency (e.g. Pierre, 2000; Nölke, 2004; Sørensen & Torfing, 2004).

It is this institutional context, and the way in which government decisions and actions are shaped, that distinguishes policy networks from other types of networks. Policy networks do not operate independently and outside the influence of government. Policy networks involve the exercise of government authority in collaboration with an active citizenry. They depart from a top-down bureaucratic approach to policy-making but nevertheless require government support if action/implementation is to occur. In this way, the state has a powerful role in shaping issue formation and framing, in empowering some interests over others, and in defining the institutional environment within and around which policy networks operate. This is a New Institutional perspective (see Immergut, 1998) that argues the state has an important but not necessarily controlling role in policy development within the new collaborative public management. The state shapes and is shaped by network formation and functioning, and the actions of the state will influence the success or otherwise of networks in policy development. Moreover, given the resources of government, its relationship to civil society and its role as guardian of public interest, networks can have an important strategic role in achieving 'good' governance (Klijn & Koppenjan, 2000).

According to Sørensen and Torfing (2005) democratic performance should be a major theme in any evaluation of policy networks. Some argue that diffusion of political power and influence to civil society should facilitate wider community engagement in policy and decision-making – that is, not just 'government, profit-making firms, and non-profit private organisations to fulfil a policy function' (Linder & Rosenau, 2000: 5) but also less prominent individuals and social groups. In so doing, policy networks may gain greater democratic legitimacy and may advance new forms of democracy. Others argue for a further strengthening of the existing political–constitutional framework of representative democracy 'as a last bastion of control which is committed to the public interest' (Schneider, 2005: 10). Others propose a combined approach of parliamentary and extra-parliamentary democracy. According to the latter view, policy networks do not replace but rather complement representative democracy.

To be able to reconcile the above views it is a necessity to better understand policy-making and the role played by non-government actors. In the tourism policy-making literature thus far, the relational constellations of policy actors and public, private and non-profit stakeholders' influence on tourism policy-making processes and outcomes has received patchy attention.

Good Governance and the Policy Network Challenge

'Good political governance', in the sense of more democratic, transparent and legitimate ways of governing, requires an effective political framework conducive to achieving shared goals and responsible decision-making. It requires an efficient state administration and a strong civil society with the capacity to engage in constructive dialogue and problem-solving (Hirst, 2000; Marsh, 2002). As a concept, it is closely aligned with the objectives of the third way politics and the shift to neoliberal public management because it embodies new relationships and organisational structures between civil society, business and government (Pierre, 2000; White, 2001). There has been much attention placed on framing good governance, with eight characteristics emerging in the literature. Table 6.1 outlines these characteristics and, drawing from the literature, identifies some of the challenges in achieving these overall principles. While not intended to be an exhaustive survey of the good governance literature – this is outside the scope of this chapter – this table does illustrate considerable tensions between the ideal and the implementation of good governance, third way politics and the management of networks. But are these same problems present in tourism policy? Do tourism networks promote good governance or have they fostered factionalised policy-making? In what follows, we examine the fragmented landscape of tourism policy network research to answer these questions.

Tourism Policy Networks Research

In tourism, policy network research is generally fragmented; insights are local and consolidated understandings have yet to emerge. This fragmentation results from unclear conceptions of policy networks and their point of difference from other types of networks. It is important to clarify this situation before any attempt to make sense of tourism policy network research. An examination of the broad tourism network literature and associated case studies reveals there are three broad issue streams:

Table 6.1 Principles of good governance

Principles	*Explanation*	*Challenges*	*References*
Participation	Participation by diverse groups/individuals Freedom of association and expression	Not all groups and individuals have equal capacity to participate in policy-making	Marsh (2002) Thompson & Pforr (2005) Agger & Löfgren (2006)
Rule of law	Fair legal frameworks Protection of human rights Independent judiciary and incorruptible police force	Unless networks have the support of bureaucracy and judiciary, network decisions and action may not be legitimated	Agger & Löfgren (2006) March & Olsen (1995)
Transparency	Decisions follow rules and guidelines Information is freely available	Informal networks may not have clear rules and member-ship guidelines and stacking may be a problem Decisions may not be representative of the network membership	Agger & Löfgren (2006) Sørensen & Torfing (2004)
Responsiveness	Institutions serve stakeholders in a timely manner	Network members may have varied capacity/expertise to respond to all problems	White (2001)
Consensus oriented	Mediation of different interests to reach agreement on what is in the best interests of the whole community	A possible tension exists between the influence of the executive authority and majority coalitions	Marsh (2002) Pforr (2004)

Continued on next page

Table 6.1 *continued*

Principles	Explanation	Challenges	References
Equity and inclusiveness	Participatory processes free from exclusion Empowering minorities to participate	Powerful interests can actively seek to marginalise other interests	Agger & Löfgren (2006) Sørensen & Torfing (2004) Thompson & Pforr (2005)
Effectiveness and efficiency	Institutions that make the best use of resources to meet the needs of society Sustainable use of resources	Network effectiveness may be influenced by whether government accepts its contributions as legitimate	White 2001 Agger & Löfgren (2006)
Accountability	Government and non-government interests are accountable to those who are affected by their decisions and to the public at large	Networks may not represent broad public interests but secular interests	White (2001) Agger & Löfgren (2006) Sørensen & Torfing (2004) March & Olsen (1995)

(1) Tourism networks that form and function predominantly around community issues (e.g. equity, power, influence and legitimacy).
(2) Tourism networks that form and function predominantly around economic and business related issues (e.g. business cluster development, product packaging, cooperative marketing).
(3) Tourism networks that form and function predominantly around environmental issues (e.g. land care, marine, forestry management).

Tourism policy networks are not defined by substantive issues such as those above. Instead, policy networks complement and intersect with issue-based networks and bring into focus the role of government, its relationships with business and community interests, and the effects of these relationships on policy content and government action about substantive issues. Figure 6.1 illustrates this distinction schematically. In this figure, policy networks overlay networks that form around substantive issues. At the centre of the figure, policy networks that successfully embrace and integrate multiple substantive issues (e.g. business, community and environmental issues) and effectively work with government to facilitate good tourism management are more likely to represent good tourism governance.

This does not suggest that tourism network policy research is distinct from and separate to other tourism network research. Much research into what is broadly referred to as 'community issues' has circled around, complemented and even intersected with network theory, exploring issues such as equity, power, influence and legitimacy. This literature includes studies of collaboration, stakeholder relationships and transactions (e.g. Palmer, 1996; Bramwell & Lane, 2000; Vernon *et al.*, 2005; Yüksel *et al.*, 2005), politics and power relations (e.g. Reed, 1997; Yüksel *et al.*, 2005), and inter-organisational relationships (e.g. Selin & Beason, 1991; Yüksel *et al.*, 2005). These interrogations conclude with observations about a particular issue or concern, such as a

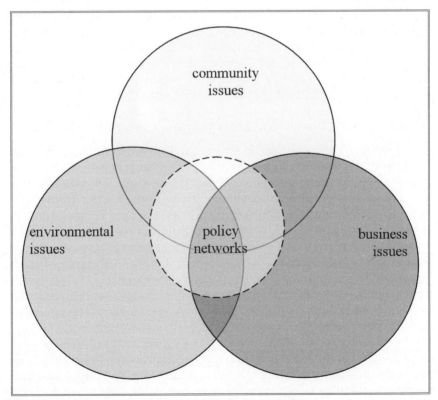

Figure 6.1 Policy networks and issue based networks

network's capacity for knowledge sharing and innovation (e.g. Pavlovich, 2001; Pforr & Megerle, 2006), or the nature of ties between private interests and political-bureaucratic actors and influence on policy content (e.g. Pforr, 2002, 2006a). In the tourism business networks literature, network theory has been discussed in terms of its effect on marketing synergies and product packaging, complementarity and cluster development (e.g. Tinsley & Lynch, 2001; Buhalis & Molinaroli, 2003; Novelli *et al.*, 2006; Pforr, 2006b). In this stream of research, the role of government receives varying attention. Research that considers the way business networks can usefully collaborate with and use the resources and power of government is generally located closer towards the centre of Figure 6.1.

The weakness in much of tourism network research is that it only superficially considers the role of government, and the effect of networks on policy content and government action/inaction. However, to be effective and achieve their goals, it is unlikely that business, environmental or community networks can operate outside of and independent to government. The remedy to bind tourism network research, therefore, is to take a more integrated approach that recognises the role of governments in empowering and legitimising networks, and that considers the resources and the unique position that governments hold in fostering tourism governance. This is the essence of the New Institutional approach previously discussed.

Good Governance in Tourism and Relevance of Networks

Yüksel *et al.* (2005) explore the devolution of tourism management in Turkey from centralised to local levels, and the blurring of lines between public and private sector responsibilities. They examine the effects of this decentralisation process on accountability, legitimacy and effectiveness, concluding that networks can exclude or marginalise some interests, that accountability is diluted and that resultant policy can be rejected as illegitimate. In an exploration of collaborative practice that does not explicitly examine networks, Vernon *et al.* (2005) find that the public sector has an important leadership role, but that the private sector also needs to accept responsibility and move away from a culture of dependency to self-reliance; a finding reminiscent of Giddens' idea that there be 'no rights without responsibilities'.

Most research consistently echoes that good tourism management requires collaborative structures and practices that allow a balance between top-down and bottom-up approaches to tourism policy formulation and implementation. Good tourism management is intimately tied to good governance (Palmer, 1996; Manning, 1998; Vernon *et al.*, 2005; Yüksel *et al.*

2005). Real benefits can be gained from applying the good governance and third way politics principles to tourism network formation and management. Our point is that networks provide a bridge between the substantive principles of good governance and third way politics and the implementation of good, responsive, transparent, accountable management of tourism in the public interest.

In what follows, we draw together the existing research on tourism policy networks, comparing their application and performance against the principles of good governance identified in Table 6.1. We examine only that research which specifically refers to the structure and operation of tourism policy networks – whatever the substantive issues of that network may be – and we collate the implications for good governance. Studies that specifically examine network structure, functions and outcomes, that consider the role of government in policy-making, and that refer to tourism policy outcomes are examined. Our objectives are, firstly, to progress an understanding of the practical benefits and disadvantages of networks from existing case study research, and secondly, to highlight the strengths and weakness of existing research and provide direction for future tourism policy network research.

Table 6.2 summarises this research and highlights implications for good governance. It suggests that while networks hold promise in achieving good governance, in practice there are many critical issues. In fact, many or all of the principles of good governance can be compromised by networks if their formation and management is poorly executed. Moreover, third way politics principles may not be furthered by the use of networks. The literature outlined in Table 6.2 suggests that networks can constrain democratic participation, transparency, inclusiveness, equity and accountability.

Discussion

These findings highlight the importance of network management for good governance. The good governance project requires critical, transparent and entrepreneurial management approaches that incorporate communication strategies, positive transactions between actors and agencies, shared goals, organisational structures that facilitate cooperation and shared learning. For tourism policy-makers interested in harnessing the potential of networks and pursuing the good governance/democratic project, it is not enough to be guided by the broad strategies and principles of good governance. This research has highlighted five dimensions of tourism policy network management that are required to achieve good governance and third way politics principles:

Table 6.2 Tourism policy network research and governance

Governance Principle	Governance Issues in Tourism Policy-making	Case Studies in Tourism
Participation	Some participants can be marginalised/excluded in network dialogues and decision-making Participants can represent narrow interests	Pavlovich (2001) Dredge (2006a, 2006b) Pforr (2006b)
Rule of law	Networked collaboration not followed up by formal policy is less effective Participation of government is required for networks to achieve their goals	Palmer (1996) Lawrence (2005) Dredge (2006b)
Transparency	Elites can have powerful influence on final policy outcomes and government action Informal networks can work to influence government outside established, transparent policy-making procedures	Pforr (2002; 2006a)
Responsiveness	Network less effective if there is no agreement on issues and priorities Issues prioritised differently / no agreement on what is important Lack of network cohesiveness and conflict can make political decision-making too difficult; political indecision renders network ineffective and illegitimate	Lawrence (2005) Pforr (2006b)

Table 6.2 *continued*

Governance Principle	Governance Issues in Tourism Policy-making	Case Studies in Tourism
Consensus-oriented	Parochial conflict can exist between participants making consensus difficult Strong consensus over common purpose facilitates information sharing Conflict may emerge between state and community interests where state not seen to be acting in the interests of the community Participants may not be willing to accept others' views delegitimate the network	Dredge (2006a) Pforr & Megerle (2006) Pavlovich (2001) Lawrence (2005) Dredge (2006a, 2006b)
Equity and inclusiveness	Network not seen to be representing all interests and decisions/solutions are then not legitimated by the remainder of the policy network Networks can promote shared values and cross-cultural understandings Collective sanctions can work against those not sharing the network's common interests	Palmer (1996) Pforr (2006a) Pforr & Megerle (2006) Pavlovich (2001)
Effectiveness and efficiency	Networks can enable clusters of businesses to participate in government initiatives/programs Accountability reached when there is development of both internal and external relationships and a 'seamless' organisation is achieved	Novelli, Schmitz & Spencer (2006) Pforr & Megerle (2006) Palmer (1996)

Continued on next page

Table 6.2 *continued*

Governance Principle	Governance Issues in Tourism Policy-making	Case Studies in Tourism
Accountability	Prominence of political and industry players in tourism networks observed Tourism policy communities characterised by state clienteleism structured around industry interests	Pforr (2002; 2006a)

(1) *Defining networks.* Networks are difficult to define. They are usually defined in terms of clusters of actors and agencies engaged in dialogues over substantive environmental, community and business issues. They can be formal or informal, and network players move in and out of active engagement as issues are identified and dealt with. As a result, they are everywhere, operating over different spatial scales, over time and across interconnected issues. Good network management acknowledges the dynamic and temporal nature of networks, keeping a long-term vision and awareness of these interconnections. Defining networks strategically will enhance the democratic and participative nature of tourism policy and facilitate inclusiveness.

(2) *Focus and structure of networks.* Network structures and relational characteristics are important determinants of a network's capacity to achieve good governance. The role of tourism policy network managers is to foster transparent collaborative structures and actor relations. Networks may have sub-networks based around interconnected issues across which communication strategies work to foster information sharing and knowledge building.

(3) *Role of the state.* A critical point in defining policy networks as opposed to issue networks is that of the role of government. Tourism policy-makers using networks to achieve good governance should carefully consider the role of government, and its capacity to influence networks' formation and management. Governments have a role in facilitating networks to enhance both government and network effectiveness and accountability to the public interest.

(4) *Episodes and continuity.* Tourism policy-making is a continuous, transformational exercise in which networks of actors move in and out of participation. Network managers must recognise the contribution of particular episodes of collective action over temporal issues, while at the same time steer the longitudinal development of the network over time. This requires that network managers balance episodes and continuity to achieve responsiveness to short and long-term interests.

(5) *Integrating macro and micro influences.* Varying attention is often placed on the interplay between micro and macro influences that shape the evolution of networks over time and across space. These influences are manifested at different spatial scales and may work to increase or decrease network participation at particular times. Network management requires an approach that balances the extenuating circumstances of critical external events against the internal routine challenges of local stakeholders.

Conclusion

This chapter has explored the use of network theory in tourism policy research and practice, and the resultant implications for 'good' tourism governance. Our objectives were (1) to progress an understanding of tourism policy network research derived from existing case study research; and (2) to highlight the strengths and weakness of research and its implications for tourism policy network management.

With respect to the former objective, tourism policy networks offer potential in achieving good tourism governance. The existing research into tourism policy networks is scant and fragmented. As a result, it is possibly still too early to draw conclusive insights into whether policy networks can achieve good governance and foster third way principles. With respect to the second objective, a policy network approach provides opportunities for more participative, democratic forms of tourism policy-making. However, where poorly managed, policy networks can also impede transparency, accountability, consensus, equity and inclusiveness. We have argued for greater emphasis to be placed on the role of government in shaping and managing networks, how government empowers/disempowers networks, and legitimates/delegitimates networks. Much of the fragmented research around tourism policy networks, collaboration, power and influence, tourism organisations and stakeholder interactions and transactions can be bought together by bringing the state back into network research, rather than treating it as an *ex-post* implication.

On this basis we identified a number of dimensions for tourism policy network management that complement and extend the principles of good governance and the third way democratic project in tourism. These dimensions must be tempered with understandings derived from practice and experience. The management of networks to achieve good governance requires a strategy based on insight. Careful observation and grounded research combined with normative principles should inform network management.

Finally, it is well beyond the scope of this chapter to place all research that circles around, complements and intersects with network theory into a framework that can guide good governance network management. However, it is possible to identify bodies of research that complement the network approach and that enhance understandings of how networks of actors and their relational characteristics can achieve good governance and the democratic project. These complementary bodies of research include:

- Studies of collaboration (collaboration theory).
- Studies of stakeholder relationships (stakeholder and transaction theory).
- Studies of politics, power and influence.
- Studies of tourism organisations.

Future research should surely seek to integrate and bind these lines of research with that of tourism policy networks in an effort to move tourism network theory and management practice beyond its current embryonic stage.

Chapter 7

Network Governance and Connectivity: A Case Study

KATHRYN PAVLOVICH

Introduction

How are networks governed? What are the coordination processes that underlie networks? Where does leadership reside in networks of interdependent organisations? These are a few of the key questions that need to be examined if a deep appreciation of the networked economy is to be understood. Unlike hierarchies, with their apparently clear lines of accountability and responsibility, networks are characterised by a complex array of interconnections that occur in a seemingly random manner. How then can we make sense of how resources are organised, and how can these connections be managed for value to be created?

These questions are pertinent to understanding the organisation of tourism destinations, with their groupings of related organisations within a geographic proximity forming networks of interdependence. Three significant features characterise these systems. First, destinations that attract large visitor numbers have a diversity of complementary firms, forming what Leiper (1990) describes as partially industrialised systems. Alongside the core attractions that pull visitors to a place are service activities such as food, accommodation and transport. Comprehensive destinations also have a variety of additional secondary attractions that add value to the destination stay, and have strong interactions with public and governmental agencies. This diversity adds complexity and texture to the network through its specialisation of niche products. The second feature is the 'comings and goings' among visitors and suppliers. Greffe (1994) argues that this process of movement from partner to partner necessitates the organisation of supply in loosely interdependent networks in order to provide the tourist experience. These 'comings and goings' can be horizontal as supplier interacts with supplier; vertical as a supplier connects with government, research institutes, or industry associations, and diagonal through cross-industry interaction. These 'comings and goings' also include the tourist visitors as they connect and pass information to each other (horizontal), to other organisations (vertical) and to other industrial sectors (diagonal). Thus, they are the connections that link different nodes in a network, a structural coupling. Unlike a hierarchical organisation, these 'comings and goings' occur in any form, with no beginning and no end point. Thus, these networks are seemingly chaotic and complex in form. Yet this very interconnectedness and complexity creates the third and final feature, a network macroculture. This macroculture is formed by the structural coupling that emerges from the 'comings and goings', and these interrelationships create a system of shared assumptions and values that form the essence of the destination (Jones *et al.*, 1997). Thus, each destination has a uniqueness that has developed over time, and this uniqueness threads though and connects the interdependent organisations. This may be physical in nature (Grand Canyon in Arizona), built structures (pyramids in Egypt; Silicon Valley) or culture (Munich beer festival, indigenous cultures). Significantly, it is the interaction among the organisations that forms the macroculture, with structural patterns of embeddedness that emerge from this process.

Central to networks then, is the nature of the connectivity that emerges from the structural coupling (comings and goings) within the relational system. This places the movement that underlies these connections at the centre of inquiry. The research question to be examined, then, is 'how are networks governed in this way?', with this chapter exploring this question through a case study. One hundred years of destination evolution and transformation is described, with the evolving network destination illustrating how coordination occurs. These data help to address what McPherson *et al.* (2001) and Powell *et al.* (2005) claim as limited longitudinal analysis of network evolution and transformation. Prefacing this is a discussion of current literature on network governance.

Network Governance

Conventional approaches to network organisation are examined through network theory, with its focus on nodes and connections. Premises underlying network theory include its three elements of network coordination: structural, relational, and embeddedness. The structural approach examines the architectural patterns of the network through the density of interconnection, and posits that networks with denser connections have quicker information flows (Brass & Burkhardt, 1992), greater access to resources (Powell *et al.*, 1996), lower transaction costs (Freeman, 1984) and higher certainty. Simultaneously, however, there are more constraints placed on action due to third-party transitivity (Granovetter, 1992; Uzzi, 1996). Much of the research from this perspective, such as that by Krackhardt (1992), examines the power relationships that exist in an actor's social networks. Also related to the structural approach is Burt's (1992) identification of 'structural holes' (the gaps between unlinked networks) that are the source of entrepreneurial opportunity.

The second element, the relational aspect, explores the nature of the nodal connections within the network, and focuses on positional aspects of centrality (Freeman, 1984), and the qualities and intensity of the connections through strong and weak ties (Granovetter, 1973). A central position within a network proposes increased and faster access to information and resources (Brass & Burkhardt, 1992), while a more peripheral position reduces this market intelligence. Rowley (1997) has provided a matrix for understanding organisational action within network structures which bridges the centrality/density gap. In this he claims a firm will take on a commander, subordinate, compromiser, or subordinate position depending on where it fits on the matrix. Finally, Granovetter's (1973, 1992) ties argument signals the nature of the relationship – whether it is strong, intense and trustworthy, or whether it is more transactional.

The final element, that of embeddedness, refers to the stockpiling of reciprocal obligations through social relations that act as the glue that binds interests together in patterns and webs of integration (Gulati, 1998; Uzzi, 1998). Granovetter (1992: 53) posits that embeddedness 'refers to the fact the economic action and outcomes ... are affected by actors' dyadic (pairwise) relations *and* by the structure of the overall network of relations'. Thus, the aggregated patterns that emerge from this mutual interconnectivity act to build social cohesion, with these linkages facilitating the flow of information and common understandings across boundaries. These benefits, then, accumulate for the network as a system through the interconnected ties. Uzzi (1997: 134–5) also declares that 'the longer an actor has made embedded contacts within their present and past networks, the more the benefits of embedded ties can be "stockpiled" for future needs'. This acknowledges not only how a macroculture is formed over time in a network, but also how obligation forms through social ties.

Thus, it is acknowledged that the network structure is the vehicle for knowledge creation processes. Dyer and Nobeoka (2000) argue that Toyota's ability to manage its knowledge-creation processes lies with its network of suppliers. These authors suggest that Toyota's accumulated knowledge is created within the networked group though the establishment of rules for entry and participation, and they illustrate how this accumulated knowledge provides competitive advantage for the network. Kogut (2000) confirms this premise, illustrating how compositions of knowledge remain the property of the network, with individual firms not having independent ownership or access to these knowledge bases. In the tourism literature too, there is substantial description of inter-organisational collaboration within tourism destinations reliant on sustainable ecotourism practices (e.g. Jamal & Getz, 1995; Medeiros de Araujo & Bramwell, 1999). Each of these narratives depicts multi-sector collaboration as a network property. This is strategically significant in that it acknowledges that the network configuration possesses its own structural form that holds the knowledge repository of the network. Thus, the relationships between organisations become critical for knowledge creation and information sharing.

It is through these three elements (structural, relational, and embeddedness) that we get a sense of how resources are coordinated within networks. However, a fundamental premise of networks is their value-adding properties, and because of the static nature of network theory, these more productive processes are not apparent. Indeed, network theory posits a staticness that is inherently a paradox, because networks are fluid structures based upon movement and change. The approach argued here is that it is the nature of the connections between nodes (the organisations) that is the critical dimension in understanding how organisation occurs in interdependent contexts. What is missing from the literature is how the movement and flow through the structural coupling in the network can be a source of value creation. This helps explain what Pentland (1999) refers to as the hiddenness of generative organising. In this way, we are able to more deeply understand how knowledge creation processes act as a network governance mechanism. The following case data illustrate these dynamic processes.

Method

The case study is set within an 'icon' tourism destination in New Zealand, the Waitomo Caves. This isolated rural village is located within a limestone karst landscape, and is famous for its tourism attraction, the Glowworm Cave. Unique to this cave is its fauna, the glowworms, which twinkle like millions of stars in the night within the underground system. The 40 minute tourism experience consists of a passive 200 metre amble through grand and imposing limestone formations within the cave, followed by a mystical river float amongst a sky of twinkling glowworms, each one no bigger than a mosquito, which inhabit the narrow chambers of the caves. Five hundred thousand people visit this village each year, in a country of only four million people.

The Glowworm Cave has been a major tourism attraction for more than 100 years, and Waitomo has historically been a single-attraction destination site with a heavy dependence on the day-trip coach market. However, a change in tourism patterns during the 1980s witnessed a growing interest in free and independent travel (FIT) by the youth adventure-seeking market. This market created a new opportunity for tourism in Waitomo with smaller specialist firms emerging to provide caving adventures. So today, two distinct production systems operate. The core tourism product, the Glowworm Cave, remains an important feature particularly within packaged coach tours of New Zealand and, because of its short stay, continues to operate independently within the Waitomo destination. The caving adventures, however, require a different type of production system, with symbiotic interdependencies forming among the primary and support tourism activities – transport, food and overnight accommodation providers. This local production network system, then, characterises a small, but dense network, with complex interorganisational dependence on both the ecological environment and the commercial market. This unique context enables illustration and understanding of governance, as these aggregated patterns have impacted on the development of the destination network.

Case data were collected over a ten year period, 1996–2006, through a mix of formal interviews, archival data, informal conversations and personal observations. From the transcribed texts, the data were analysed in a manner consistent with interpretative research methods, through an iterative process of thematic analysis (Miles & Huberman, 1994; Locke & Golden-Biddle, 1997). This enabled an evolutionary and transformational perspective of the tourism destination to be crafted so that the structural dimensions of network governance could then be reconstructed from both the textual narratives and research-oriented observations and conversations. To illustrate this evolutionary structural connectivity, a visual diagram of nodal interconnection was produced in accordance with the format used by network analysts (Burt, 1992; Krackhardt, 1992; Madhavan *et al.*, 1998). The insights from this analysis are included in the following discussion, which firstly reconstructs case data and then identifies generative organising principles that expand upon our understandings of network governance through movement and flow.

Case Data

The following data are presented within four sections. The first describes early development of the destination, and its history of exploitation; the second discusses the formation of a networked destination which sets the context for network-based capabilities to be developed, as evident in section three; while the final section discusses the destination's current state as a learning network. As noted above, this development of a learning network is illustrated through a focus on structural couplings as movement and flow.

1887–1986: Evolution of the destination network – One hundred years of exploitation

Tourism in Waitomo began in 1887, after local Maori chief, Tane Tinorau, showed the Glowworm Cave to his friend, British surveyor Fred Mace. Mace immediately notified the New Zealand state government of its existence, as the previous year an eruption of Mt Tarawera had destroyed the world famous Pink and White Terraces in nearby Rotorua, claimed to be one of the great wonders of the modern world. The Government immediately recognised that the Glowworm Cave could be an important alternative tourism destination, and in 1903 it was nationalised under State management, supposedly to ensure its protection and conservation for future generations (Arrell, 1984).

By 1910, an embryonic destination had begun to develop. Alongside the core attraction were transport (a regular coach service, a blacksmith's shop and stables), accommodation (a private boarding establishment and the Government hostel which still forms the old wing of the present Waitomo Hotel), and support activities (the general store). The Government (through the state Tourist Department) retained control of the emergent destination with its governance of the Glowworm Cave and the accommodation sector. Although simple, a complementary production system had emerged to provide a comprehensive visitor experience.

However, over the following 70 years, there was minimal investment in the destination and little development and growth occurred. The lack of structural change can be seen in Figure 7.1 which compares the nodal structure of the destination in 1910 and 1986.

It is clear from Figure 7.1 that there was little fluidity and adaptability in this destination over the 100 year history. Indeed, the structural history indicates meagre growth in enterprise activity during this time, despite Waitomo being a major tourism destination. While the central organisation, the government-owned Tourism Hotel Corporation (THC) had impressive links with national tourism sectors, there were no additional linkages with other organisations within Waitomo. Such restricted access, noted by Jones *et al.* (1996), resulted in status maximisation as the central firm interacted only with firms of equal status. In this case, the ties developed by THC were

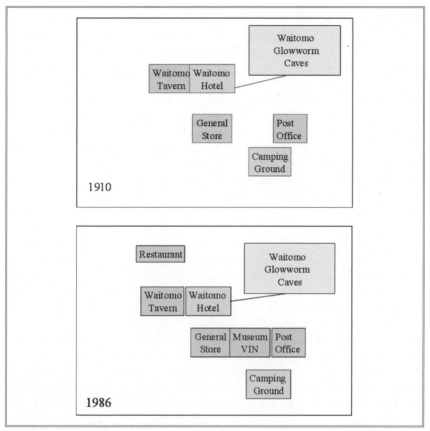

Figure 7.1 Comparison of the nodal structure in Waitomo: 1910 and 1986

outside the destination, with benefits accruing only to the external status organisations. Through this strategic reduction, the THC separated itself from the wider community within which it was embedded, thus denying the mutual learning opportunities that can evolve from reciprocity. The macroculture of this network was based upon constraint, separation and inertia, which contributed to low adaptability, resulting in a stagnation of destination vibrancy, as evident in the single-attraction destination history. This confirms the lack of 'comings and goings' among different suppliers in the destination. Thus, movement and information transfer for learning from the network's structural coupling were limited. Governance of the destination, then, through its absence of structural connections within the network, was organised as an isolated node with limited connectivity.

Radical change and evolution of the tourism network: 1987 onwards

The dramatic growth in international visitors to New Zealand during the late 1980s, the inert state of the THC and the growing frustrations within the Waitomo Caves community all crystallised into major upheavals over the following decade. First, the land on which the Glowworm Cave was located was returned to the indigenous Maori people in 1990. This meant that for the first time since the land confiscation in 1903, Maori were included in strategic decision-making within the tourism destination. Second, the State (the THC) sold its interest in the Caves and Hotel, with the cave operations being leased by Tourism Holdings Limited in 1996.

Finally, adventure tourism arrived. With its unique limestone karst formations, flora and fauna, the Waitomo Caves region offered splendid opportunities for participation in leisure, environmental and aesthetic experiences. In 1987, Blackwater Rafting (BWR) was formed by two local residents who foresaw the potential in commercialising the 'laundry trip', a fun trip that recreational cavers used to clean their clothes after caving. Suited out in wetsuits, the rafters and guide walked through native bush reserve and, after entering the cave, floated down the river stream in tyre tubes amidst the darkness and glowworms of the neighbouring capacious Ruakuri Cave. Shortly afterwards, a second adventure enterprise, Waitomo Adventures (WA), was initiated by a non-local person, who came to Waitomo expressly to develop the Lost World adventure tourism business, which involved a 100-metre abseil into the cave shaft. The third adventure operation, Waitomo Down Under (WDU), began operating in 1992. With its owners being part of the Maori community, this organisation provided a uniquely indigenous perspective.

These new ventures provided more authentic and compelling personal experiences than that of the passive Glowworm Cave, and their arrival was described as changing the nature of tourism in Waitomo as they brought back overnight stays. By the year 2000, 50,000 people came to Waitomo each year to participate in adventure caving. Numerous support activities were developed to complement the caving activities: backpacker lodges, horse trekking, a guest lodge, canoe caving, an agricultural pioneer show and jet-boating, alongside the original camping-ground and general store. While the population of this village remained a constant 300 people, 200 full-time equivalent jobs were now located within Waitomo. The core Glowworm Cave product remained the icon attraction, but the adventure caving products created a new market complementing the primary activity and adding supporting activities to extend the value of the destination. The nodal

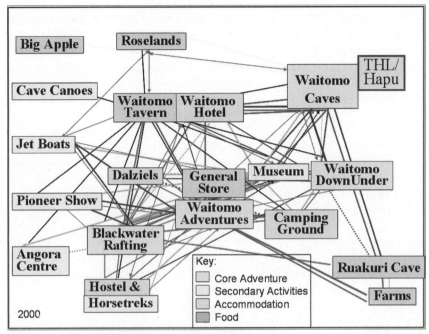

Figure 7.2 Nodal structure in Waitomo, 2000

structure of the network in the year 2000, as illustrated in Figure 7.2, includes the complexity of interconnections in the network.

During fifteen years (1987–2002), significant structural change has occurred in Waitomo, and Figure 7.2 depicts a more complex nodal configuration of tourism supply than the 1986 structure presented in Figure 7.1. Furthermore, for the first time, exchange relationships began developing in the destination as visitors were passed from one supplier to the next, and an ecosystem of interdependence was formed (to be discussed in the next section).

However, the embedded memory of the destination remembers the restricted access from the THC governance that has resulted in little knowledge accumulation over this time (Jones *et al.*, 1997). Thus, the exploitative history of the THC continued to impact in Waitomo throughout the 1990s. Following the sale of THC, the Glowworm Cave continued its simple routines with little information sharing and absence of knowledge creation processes. Its routines mainly involved profit retention and focused largely on guiding visitors through the cave. Administrative technologies were limited to telephones and faxes until 1996, and while a research institute was contracted to assist in the measurement of cave hydrology, there was no insistence on the analysis and interpretation of results until the mid-1990s. It was not until Tourism Holdings acquired the lease of the Glowworm Cave in 1996 that specialist systems and processes were developed around cave monitoring processes involving cave hydrology, flora and fauna management, limestone conservation, and administrative and educational systems.

The preceding section illustrated how limitation lingered as the central organisation was unable to initiate leadership within the emergent network. There was little movement and flow of specialised information sharing from the central organisation to the broader network. Indeed, this confirms the findings of Powell *et al.* (1996) whereby older, less linked organisations are less likely to survive in the networked economy. Rather it was the newly emergent organisations, with their own portfolio of ties, that provided the bridges across many structural holes as they seized opportunities and introduced new ideas (Burt, 1992). Network governance, illustrated in the following section, occurred through the more peripheral organisations unencumbered by the previous structural constraints. It was through these 'plural pathways' (Powell *et al.*, 2005) that these connections contributed to the development of learning capabilities in the destination.

Capability building within the network

As the destination changed from a single short-stay attraction to a wider collection of adventure activities requiring overnight stays, so did the exchange of information and sharing among tourism suppliers, which has resulted in the development of network-based capabilities. Capabilities are what Eisenhardt and Martin (2000: 1107) describe as 'the antecedent organisational and strategic routines by which managers alter their resource base to generate new value-creating strategies'. When these resources integrate complementarities, they can become activity related systems that have the potential to create sustained competitive advantage. The original data analysis identified several areas of interdependent collaboration that have the potential to build network-based strategic capabilities, contributing to the stockpiling of benefits for the future leverage of the network. Three of these groupings are briefly described, noting especially the different structural configurations. Following this description, the theoretical implications will illustrate how information sharing within each forms knowledge from the movement in the structure.

Above-Ground Landcare

The first and most robust of these collaborative groups was the above-ground Waitomo Landcare Group formed in 1992. Facilitated by the Regional Council, Environment Waikato, the group included the local farming community (donating land to be fenced and replanted), tourism operators (contributing financial inputs), community members (providing voluntary labour) and other outside institutions (Waitomo District Council, Department of Conservation, and universities). In this group, there is minimal replication of similar activities; rather, there is rich differentiation and an appreciation of the different tasks, resources, and outcomes that built the land and karst management capability.

The purpose of the Landcare Group was to protect the Waitomo Caves system from sedimentation through appropriate and sustainable land management practices. Collaboration among these internal and external stakeholders included fencing off the waterways, and forest replanting programmes, primarily to improve water quality for sustaining glow-worm populations and the cave environment. The group is promoted throughout New Zealand as a 'best practice' example of a Landcare group. Its initial purpose of improving the longevity of the cave environment has been achieved, and the participation of the community has resulted in widespread consensus and acceptance of the land management practices to sustain the destination environment. Commercial operators, who assisted in funding the programme, are now required to include cave and karst conservation issues in their asset management plans, and visitors to this region are educated on these principles through tourism product delivery processes.

Adventure Risk Management

The second collaborative group involves managing the risk attached to the underground cave environment. Given that caves are dark, cold, wet and often confined spaces, the ability to manage these potentially dangerous situations requires an advanced standard of guiding to ensure client safety. This network-based capability has been developed in three ways. First, informal information exchanges among operators in Waitomo have enabled better safety practices, stemming from experience of their own organisational routines. A second mechanism has occurred through the formation of the Waitomo Caves Rescue Team, with specialist guides from the adventure caving organisations working together in emergency situations – most commonly arising from situations involving recreational cavers. Finally, the adventure caving organisations in Waitomo have been instrumental in building abseiling and caving standards throughout New

Zealand (Waitomo Standard Rope Technique – WSRT), as their systems and standards have been incorporated into institutional frameworks (New Zealand Qualifications Authority). Thus, these organisations are seen as adventure sector leaders in New Zealand.

Waitomo Caves Marketing

The third area of collaborative activity involves marketing. First, there is the Destination Waitomo Group, which involves many of the local operators. The main focus of this group is brochure development to reach domestic marketing channels, as international marketing is the focus of a government body, Tourism New Zealand. Second, the association with Tourism Holdings, licensee of the Glowworm Cave, brings many benefits to Waitomo through its centrality in the New Zealand tourism industry. Tourism Holdings has important connections with major operators such as Air New Zealand and Tourism New Zealand, meaning that much of the destination's marketing occurs through these multi-faceted external connections. One example is Waitomo's inclusion in Air New Zealand's promotion of New Zealand alongside other Tourism Holdings products.

These three network-based capabilities all illustrate connectivity as being significant in facilitating the exchange of information. The question being examined here then, is how were these knowledge capabilities formed to create value in this destination, and why did this not occur under THC governance? Following Pentland's (1999) claims that generative organising mechanisms are hidden, the role of movement and flow as an organising mechanism is now examined through a structural inquiry.

Figure 7.3 illustrates the structural coupling around these capabilities, and offers three contributions to understanding how movement and flow organise cohesive networks. First, the *heterogeneity* of activities, accommodation and food, transport and a variety of external institutions contributed to building a depth and texture to the destination structure that enabled growth to occur at each nodal point. This again illustrates how plural pathways provide richer information points, more rapid information transmission, and faster market response (Powell *et al.*, 1996). This network is in contrast to the earlier destination structure illustrated in Figure 7.1 which had little diversity in product supply. Furthermore, there were few connections between the organisations that were there. Thus, little interaction resulted in limited exchanges of information. Now, however, each of these supplier nodes has access to information.

The structure of the network facilitates the complexity of connections and the complementarity of those connections assists in movement and flow. This diversity of exchanges is based around the movement of people from organisation to organisation (e.g. activities, accommodation, food, transport); information passing from external institutions and government agencies into Waitomo; product and market innovations developing through collaborative problem-solving; and employees moving from one organisation to another, passing on information and clients and facilitating better operational practice among the organisations.

The second feature is the manner in which these connections are the vehicle for knowledge building and community development within the network. The diversity and complementarity of nodal connections enables a significant number of *structural couplings* to occur in the network. As illustrated above, these structural couplings build network-based capabilities of strategic significance that include not only economic and political organisations, but also aspects of the community within which they reside. Thus, it is this structural coupling, based on nodal diversity and connectivity, that makes the most contribution in understanding network governance. With a diversity of nodes, any number of combinations can be

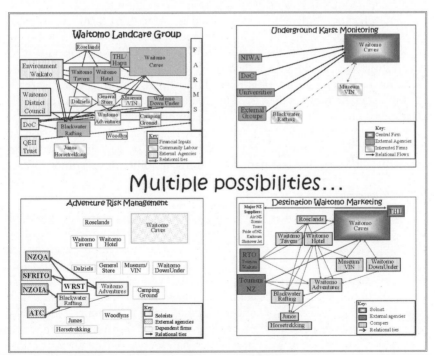

Figure 7.3 Nodal structure of network capabilities

developed for information exchanges. Indeed, the number of combinations is limited only by possibility (see Figure 7.3 for these capabilities as structural possibilities). Unlike hierarchies that organise linearly, networks have the potential to form complex layered systems from this structural coupling. This layering develops the texture and richness of knowledge-intense capabilities through the networks, limited only by an organisation's competitive or *collaborative intention*. Indeed, it is the collaborative dynamic that assists in creating more intense relationships that emerge from the structural coupling. Market-based interactions based only on pricing mechanisms do not transfer specialised information. As these capabilities illustrate, reciprocal information exchanges are based on mutual sharing of ideas, and the structural coupling occurs through task diversity as they increase a variety of knowledge repositories in the network.

Thus, knowledge-based capabilities underlie competitive advantage for the destination and are formed through more formalised collaborative endeavours that result from the 'comings and goings' between partners, employees, community and visitors. One such example is illustrated by seasonal employment, which is characterised by short-term casual relationships formed between employees. One operator said, 'Who sells our product depends on who their boyfriend is today'. In these informal ways, organisational knowledge was transferred from one activity to another, building a transferability of inter-organisational routines and systems through social connectivity. More formal approaches also occurred, with employees changing employment from one organisation to another. As they brought their old routines into a new organisation, a system of best practices began to emerge at a network level, increasing the strategic competitiveness of the destination through capability development.

2006: A learning network

The ensuing six years has seen a cohesiveness develop in the destination, with new products, alliances and infrastructure development further changing the face of Waitomo (see Figure 7.4). The most significant product development has been the re-opening of the Ruakuri Cave in 2005. Originally one of the three caves making up the Waitomo caves, the original survey boundaries were contested in the 1980s and it was discovered that a local landowner had a significant ownership stake in Ruakuri. A satisfactory outcome for its commercial operation could not be found, and the cave was closed in 1986. The last three years has seen a partnership develop between the landowner, Blackwater Rafting and Tourism Holdings to redevelop and reopen Ruakuri, at a cost of $4 million. This partnership included the sale of Blackwater Rafting to Tourism Holdings, who now have ownership of the adventure product, the newly opened Ruakuri Cave, and the operating licence for the core Glowworm Cave and Aranui Cave. This gives Tourism Holdings significant control over the core attractions in Waitomo.

Alongside this integration is the development of further secondary businesses attracted by Waitomo's success: Kiwi Paka (youth hostel) and Morepork Café, the redevelopment of the Top 10 Holiday Park, Woodlyn Park, and Spellbound. Further plans are also in place for an upmarket café, and a family restaurant and bar complex. The hotel too is beginning a multi-million dollar upgrade.

A significant expansion has been the extensive alliances, partnerships and commercial connections that have developed, both locally and externally. Figure 7.4 illustrates these governance alliances and formalised connections as they include related partnerships that extend beyond the destination context. This new stage of interconnectedness and governance consolidation is likely to bring a further level of investment, infrastructure

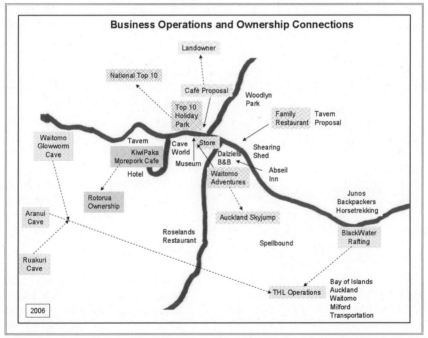

Figure 7.4 Nodal structure in Waitomo, including governance connections, 2006

development and complementarity of product offerings, impacting on the destination as a more durable tourism stay in terms of quality and comprehensiveness. Equally as important is the impact of these on the capability development in the network from the structural coupling.

Discussion and Conclusion

The key finding from this research is how the network connectivity of the last 20 years has created a comprehensive tourism destination to increase the visitor stay. This was contrasted with the absence of connectivity over the first 100 years. Yet the manner in which theorising of networks currently exists does not capture the movement and fluidity that underlies networks, seen in tourism destinations as the 'comings and goings' evident in both supply and demand.

In taking this approach, the data offer the following premises in relation to understanding network governance. First, the mapping process allows us to identify who is in the network. This structural dimension offers insights into the diversity and density of connections within the network, thus indicating *the overall health of the network*; that is, whether it is under, over or effectively connected. Coherent tourism destinations comprise diverse complementary activities (core, support and infrastructural) that form an interdependent system of relatedness. It is this nodal connectivity that underlies the ability for structural coupling to occur for knowledge creation. Second, in examining the relational aspect of networks, one can identify *how connections may better be structured in order to build value in the destination*. Remembering that an infinite number of structural couplings are possible, it is the nature of the information transmission process that becomes significant in building and developing new knowledge. Generated through connections that act as the vehicle for information transfer, reciprocal, socially construed and knowledge-intensive relationships have the potential to build vital and sustainable network contexts. In this way, it is the movement and flow emanating from the structural coupling that provides the vehicle for information transfer. As information is transmitted from node to node it is added to, and the cumulative process builds network-based capabilties that lead to strategic leverage.

The third and final element, the embeddedness component of networks, recognises how the past impacts on the future. Premises underlying embeddedness acknowledge that no organisation in a network acts independently, and the actions of one impact on others. One example of this is risk management (earlier discussed as one of Waitomo's strategic capabilities) – if an adventure firm has a serious accident in the cave, it would negatively impact on the reputation of all the organisations in Waitomo by their very geographic proximity. Another aspect of embeddedness, more positively, is that Waitomo has finally shaken off its history of exploitation by the THC, with strategic repositories of knowledge now residing in the network. One example is that Waitomo is now seen as a global centre for cave education and cave management systems. Embeddedness, then, acknowledges that the destination as a system functions not only through its individual components, but also through the connections that bind it together. Thus, underlying the interdependence premise is a mutual awareness that each organisation is implicitly reliant on the activities of others, and that they mutually construct and organise the strategic future of the destination.

Finally, this chapter has contributed to an understanding of network governance through its description of the evolution and transformation of a tourism destination over its 120 year history. Arguing that cohesive networks are organised through structural connectivity, it is the nature and intensity of connections that assist networks to grow and adapt over time. The case data illustrate how the destination has grown and evolved over the last 20 years as a learning network, contrasted with the constraint, limitation and absence of flexibility that characterised the earlier destination history. Indeed, as the network is now internally growing, learning and adapting through its complex interconnectivity, it now picks up from the 1910 state where it was an embryonic destination with many possibilities. At this stage, with multiple governmental connections, Waitomo had the potential to develop into a coherent network based on structural diversity and relational intensity. However, as previously stated, the destination was governed by a single organisation that consciously limited external connections which then stagnated destination growth.

In arguing that network governance occurs through the movement and flow of inter-organisational connectivity, this chapter implies that a new way of understanding organisation is required. Rather than placing competitiveness as the centre of organisational activity, it is reciprocity through quality connections that becomes the central mechanism driving network organisation. Thus, through understanding the role and function of structural connectivity, these 'comings and goings' so characteristic of tourism destinations, networks are more able to realise their potential value through reciprocal and collaborative exchanges.

Acknowledgements

I wish to thank Tourism Management for their permission to republish parts of the case study.

Chapter 8

The Benefits of Networks for Small and Medium Sized Tourism Enterprises

CARLOS COSTA, ZÉLIA BREDA, RUI COSTA and JOANA MIGUÉNS

Introduction

In the contemporary competitive business world enterprises have to compete globally, but must also adjust to the surrounding environment in order to become more efficient and effective locally. Networks can play a significant role in accomplishing this, as they facilitate access to knowledge, resources, markets and technologies. The literature recognises that networks have the ability to convey information and to induce innovation through knowledge exchange and shared strategies. Nonetheless, there is limited research on whether networks and clusters can be used as an innovative process to support tourism enterprises. This chapter reports on an empirical study that was carried out in Portugal, targeting sports and adventure tourism enterprises, mainly consisting of small and medium sized enterprises (SMEs). Micro and small enterprises dominate the tourism sector in Portugal and are generally located outside the main tourism routes. Despite their size, they assume an important role in local economic development, delivering an enormous contribution to the development of peripheral areas. Recent transformations in the global market have however posed a challenge to these enterprises, which now recognise the importance of cooperation with other tourism partners as a way to develop strategic positioning and capacity to operate in a competitive environment. By cooperating in the form of geographical and product-based clusters, enterprises can function as dynamic and innovative organisations. This study explores the benefits of innovative networks and partnerships, and thereby investigates how they minimize the growth constraints of enterprises, increase competitiveness, promote innovation and facilitate internationalisation.

The Emergence of Network Theory in the Tourism Industry

Networks have become a fashionable topic in the tourism literature. Despite the fact that most of the insights into the relationship between networks and business activities started to emerge in the late 1980s and early 1990s, the concept and theory of network is not recent. The origins of network theory and practice may be found in research conducted in the late 1950s and 1960s on 'social behaviour as exchange' (Homans, 1958, 1974), the 'social psychology of groups' (Thibaut & Kelly, 1959), 'exchange and power' (Blau, 1964), 'operational research in local government' (Ward, 1964), and 'inter-organisational and exchange analysis' (Levine & White, 1961; Miller, 1958; Reid, 1964).

It was nevertheless from the mid-1980s onwards that network analysis and theory began to attract the attention of growing numbers of economists, planners, sociologists, geographers, psychologists and politicians. Since then, larger numbers of academics and practitioners have become sensitive to the importance of intra- and inter-organisational environmental design to an organisation's profitability, efficiency and effectiveness as well as to employee motivation (and thus productivity).

As a result of globalisation, the vertical and horizontal integration of the industries, and the opening of the world's frontiers, competition in the tourism sector has increased significantly. Hence, businesses are becoming increasingly aware that their operation has to become more efficient and effective. Entrepreneurs are conscious that in order to compete globally, their companies have to interconnect tightly with the surrounding local environment in order to become more efficient in their operation, reduce costs, incorporate raw materials and recruit the most suitable human resources. In other words, increasing worldwide competition is pushing companies to become locally more efficient and effective.

The organisational environment is crucial for the success of an organisation because interdependence pervades organisations and is fundamental to understanding them. Individuals within a group, work groups within departments, and departments within organisations all depend upon each other. Even persons who work independently at their own job typically require others to provide information and supplies to complete their work. The links which are established between an organisation and its environment are so important that some theorists have proposed that interdependence and subsequent interaction among individuals and groups are the basis of organisations (Tjosvold, 1986: 517). Based on this argument Morgan (1988, in Knoke, 1990: 94) claims that the idea of a discrete organisation with identifiable boundaries is breaking down.

Organisations are becoming more like amorphous networks of interdependent organisations, where no element is in firm control. Even though such a network may have a focal organisation, the focal organisation is as dependent as the other organisations in the network. Interdependence is the key. Gone is the old-fashioned notion of hierarchy, in which one member (for example, the focal organisation) directs the activities of other members. In comes the notion of a network that must be managed as a system of interdependent stakeholders.

In the contemporary competitive and globalised world, the importance of networks as facilitators to access knowledge, resources, markets and technology, is critical. Network theory in the tourism sector dates back only a few years, as it has been recognised that relationships between enterprises can stimulate inter-organisational learning and knowledge exchange, resulting in qualitative and/or quantitative benefits to businesses, communities and destinations (Morrison *et al.*, 2004: 2). Therefore, several studies have been made using networking, clustering and agglomeration theories to explain the role of tourism in influencing local growth and stimulating regional development (see Braun, 2003a, 2003b; Breda *et al.*, 2006; Buhalis & Molinaroli, 2003; Costa, 1996; Dredge, 2006a; Nordin, 2003; Pavlovich, 2003; Saxena, 2005; Shih, 2005; Stokes, 2006). Despite the growing literature on the importance of networks in the tourism sector, there is limited discussion on whether networks and clusters can be used as an innovative process to support tourism enterprises and contribute to local development (see Novelli *et al.*, 2006; Sundbo *et al.*, in press). The relationship between innovation and networks is one of the most frequently discussed issues in relevant fields in economics, managerial science and sociology (Fukugawa, 2006), but seems to be a new topic in the tourism field.

The Capacity of Networks to Convey Information and to Induce Innovation

It is recognised in the literature that networks have the capacity to accelerate the exchange of information, to speed up business among entrepreneurs and also to induce innovation (Bellamy *et al.*, 1995; Burt, 1980a, 1980b, 1990; DiMaggio & Powell, 1983; Galaskiewicz & Burt, 1991; Galaskiewicz & Wasserman, 1989; Jarillo, 1993; Li, 1995; Ohmae, 1995; Osborne & Gaebler, 1992; Sweeting, 1995; Tushman, 1977). Most of the literature focusing on this matter does so on the basis of evidence showing that there is an inverse relationship between the amount of extra-organisational communication of an individual and his or her performance (Tushman, 1977: 591). Based on empirical evidence collected from several UK companies, Sweeting (1995: 91) also claims that the best innovation examples identified by participating companies were characterised by more flexible, co-operative and motivated use of people through either a process of self-help and empowerment or more enlightened and less bureaucratic management practices. The research conducted by Bolwijn and Kumpe (1990, in Sweeting, 1995: 88) also led them to conclude that innovative ability will be the defining characteristic of successful 1990s companies.

Bearing that in mind, authors such as Tushman suggest that organisations should create 'special boundaries roles', which are nothing but mechanisms (e.g., individuals, organisations, technological conveyors, etc) allowing an easier diffusion of information into organisations. The advantage of doing this seems obvious. That is, organisations may more quickly adopt new management and technological styles and thus increase their productivity, efficiency, the performance and motivation of their personnel and, generally speaking, improve their profitability and competitive advantage.

It should nevertheless be pointed out that the exchange of information is not only seen in literature directed to organisations. Bellamy *et al.* (1995) provide a useful article illustrating how organisations may also create 'special boundaries roles' with the public, in order to facilitate their access to information available within organisations. Their paper contains useful guidelines explaining how to improve the access of the public to information stored, for instance, in tourist offices (e.g., accommodation, transportation facilities, entertainment, etc). This theme is usually labelled in the literature as EIP – exchange of information with the public.

Networks are seen in the literature as organisational frameworks with great potential to facilitate diffusion of information because they strengthen connections among organisations and individuals. Galaskiewicz and Burt (1991: 89) point out that interpersonal contagion favours structural equivalence over cohesion. According to them, contagion arises from role playing among people who perform similar roles. Oliver also argues that the strengthening of the relationships among organisations leads to higher levels of isomorphism. According to her, isomorphism is the result of competitive pressures that force organisations facing the same set of environmental constraints to adopt similar characteristics relative to one another (Oliver, 1988: 542). Isomorphism then develops from the restructuring of an organisational field into an interconnected collectivity that pushes organisations towards homogeneity. In short, isomorphism is a process that emerges from the interconnectedness set up among organisations (DiMaggio & Powell, 1983).

It might, nevertheless, be thought that isomorphism may lead organisations to adopt the same operating styles, a situation which, sooner or later, may reduce their competitive and survival prospects. However, based on research conducted by other academics, Oliver argues that neither competition nor interaction will necessarily lead to reduction in the organisational field. To this she also adds that the environment is highly deterministic in shaping organisational forms and destinies; '(...) homogeneity is induced by institutional rather than competitive forces' (Oliver, 1988: 545).

The potential offered by networks for conveying information and inducing innovation may be summed up in the following main points (Burt, 1980a).

- First, research has shown that social integration is associated with early adoption of innovations (p. 329). The research conducted by Sweeting (1995) also points in the same direction.
- Second, a set of persons socially integrated within a cohesive group will react similarly to an innovation.
- Third, persons connected by intense relations will have similar attitudes towards an innovation, adopting it at approximately the same time (p. 330).
- Fourth, marginal persons at the periphery of the social structure are not normally innovative and tend to discover innovations on their own (p. 331).

Supporting the argument that networks may work as potential conveyors of information and innovation, DiMaggio and Powell (1983: 152) claim that networks create a pool of almost interchangeable individuals responsible for conveying information to other organisations and professionals, which, according to Galaskiewicz and Burt (1991: 90) makes them professional communities of multiple positions, each with its own internally reproducing beliefs and attitudes about professional work (see also Sweeting, 1995).

Innovative Process in SMEs

Considering the globalised and competitive environment in which enterprises operate, there is a growing interest in understanding the way that SMEs maintain and grow their businesses. Special attention has been given to how increased synergies and productivity, knowledge transfer and innovation take place among these enterprises. Contrary to Schumpeter's assumption that the existence of large enterprises was a prerequisite to technological change (Schumpeter, 1950), recent studies underline the economic significance of SMEs, thus leading to a growing research interest in their role in the innovation process, and even originating theories that defend the innovative advantage of SMEs.

According to Acs and Audretsch (1988), the growing activity in innovation tends to be more pre-eminent in SMEs rather than in large firms. Link and Rees (1991) also point out the higher level and significance of innovation in SMEs. One of the reasons might be the existence of interactions among SMEs and considerable research activities between universities and these firms (Acs *et al.*, 1994; Jaffe, 1989; Link & Rees, 1991). Even though large firms are more active in engaging in research partnerships, SMEs are able to use their relationships more effectively to stimulate internal research and development (R&D) activities at a higher level. SMEs thus tend to compensate for the lack of R&D through spillovers and spin-off effects (Acs & Audretsch, 1993; Jaffe, 1989; Mytelka, 2002).

Learning between firms and suppliers (Lorenzoni & Lipparini, 1999) and the exchange of knowledge between universities and technological companies (Kreiner & Schulz, 1993) represent highly informal innovation networks, which are very important but hard to measure (Freeman, 1991). These informal networks, by exchanging knowledge, can be an essential factor for the development and diffusion of know-how, representing the innovation process front line (Smith & Reinertsen, 1991). The innovation is promoted through efficient transfer of codified knowledge (Fukugawa, 2006).

In a global economy with increased pressure on SMEs, partnerships and networks of enterprises are inevitable. Innovation, cooperation and collaboration are essential to achieve competitiveness, and these competitive advantages can be found at the local level: knowledge exchange and relationships among stakeholders (Smeral, 1998). Considering that through a cluster, a group of SMEs can compete globally by cooperating locally, networks and clusters in tourism have experienced a dramatic growth, bringing benefits such as flexibility, a share of valuable marketing information, innovation, opportunity to enter other networks and clusters on a national level and across borders, resource development and knowledge transfer between stakeholders (Novelli *et al.*, 2006: 1143).

The Structure of the Tourism Sector in Portugal

The tourism sector is one of the fastest growing industries in Portugal, representing an important contribution to the balance of payments, investment, income and employment generation. Directly and indirectly, tourism provides jobs for about half a million people and accounts for approximately 11% of GDP. Despite its enormous potential, Portugal still has a long way to go with regard to tourism. It is among the top 20 destinations in the world but is losing competitiveness and lagging behind other European destinations offering similar products (see Figure 8.1). In 1985, it ranked 15th in the world, dropping to 19th place in 2004. In terms of earnings from tourism, in 2004 Portugal was ranked 22nd, dropping from 20th in 2003. Annual growth rates for Portuguese tourism – both in terms of arrivals and receipts – are positive, but have fluctuated over the years, with arrivals always outpacing receipts. Statistics suggest that Portugal has a long way to go in terms of tourism earnings. This situation is evidence that Portugal has been a relatively inexpensive destination, or it may suggest that Portugal is attracting the lower end of the global tourism market.

It is also important to underline the high dependency of Portuguese tourism on a few European countries, as well as the concentration of tourists in certain regions. There are three main tourist destinations in Portugal – the Algarve, Madeira and the Lisbon area – which account for 77% of the total bed nights in the country, in part because of the high concentration of tourist facilities and also because these regions are sun-sea-sand-oriented.

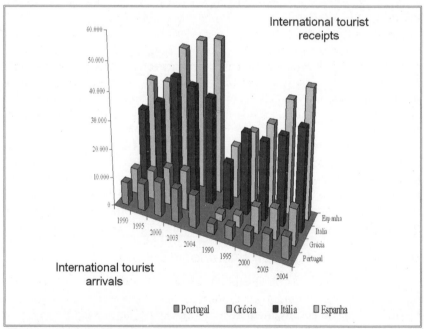

Figure 8.1 International tourism arrivals and receipts in Portugal and other south-western European countries. (Note: data not available for international tourism arrivals in Greece in 2004.)

The Algarve has become the major Portuguese tourist destination but its significance in terms of bed nights has decreased gradually. The Lisbon region, with a prevalence of MICE (Meetings, Incentives, Conventions and Events) tourism, has become the largest region in terms of receipts, which shows that Portuguese tourism is undergoing major changes by attracting new markets, by diversifying products and by changing its economic structure. Nonetheless, the Portuguese tourism industry still suffers from fundamental problems. Perhaps the lack of a well-designed tourism strategy is responsible for the inability of the Portuguese tourism market to achieve its global potential (Yasin *et al.*, 2003).

Portugal needs to change its image, which is heavily associated with sun and sea, by encouraging the development of new products, by attracting more affluent tourists and by attempting to reduce regional imbalances. Although relatively small, Portugal offers an interesting tourist diversity, with unique landscapes, environmental features, culture and traditions which are now being revalued in post-modern societies. Therefore, nature and culture-based products should be encouraged, especially by SMEs since they play an important role in the economy and in the development of destinations. The important role that SMEs can play in the innovation process, as well as their impact in employment generation, makes them more effective in stimulating regional development (Armstrong & Taylor, 2000).

SMEs, especially micro-scale family businesses, dominate the tourism sector in Portugal. These generally tend to be located outside the main tourist routes, thus contributing to the economic development of peripheral areas. Globalisation poses a big challenge to these enterprises that must strive to be competitive. One approach, which helps enterprises and destinations to cope with their internal limitations and challenges posed by the transformations occurring within the global market, is the development of networks and partnerships. By cooperating with other tourism industry partners, these enterprises are able to develop strategic positioning, extend competences, identify opportunities and threats, and build up capacity to operate in a competitive tourism environment. According to Sundbo *et al.* (in press), destination building based on large-scale tourism firms may sustain innovative and competitive destinations, but small firms result in highly innovative destinations as well.

Methodology

SMEs clearly assume an important role in economic development, but their competitiveness, growth and profitability are affected by a number of factors. To minimise growth constraints they must strive for the establishment of strong partnerships and networks with private sector operators and with public sector entities. Networks with a common strategic orientation can evolve into geographical or product-based clusters. In order to analyse the performance of a tourism product-based cluster in Portugal, a study was conducted targeting sports and adventure tourism enterprises (Costa, 2005). The study had the following main objectives:

- To characterise the business structure of enterprises operating in this tourism sub-sector.
- To determine their business area and identify their main objectives, strategies and constraints.
- To present the main advantages of the existence of a tourism network in their business area so as to increase competitiveness, to promote innovation and to facilitate internationalisation.

The study is based on a mail questionnaire sent to the managers/owners of all firms belonging to this specific tourism sub-sector, drawn from the official registry of the National Tourism Board. In total, 237 establishments were contacted for the survey and, of them, 107 returned questionnaires, which constitutes a general response rate of 44%. Responses were collected between October 2004 and January 2005. The questionnaire was divided into six main sections (see Figure 8.2): the first relates to the general characterisation of firms; followed by a section in which respondents are asked about their objectives and strategies, so as to characterise their business and entrepreneurial structure. The third section is about the way these enterprises relate and interact with each other and with the environment that surrounds them. The existence of business relationships, with more or less ties, can be important for the establishment of agreements and joint efforts for attaining common goals. For this reason, an analysis of interactions between these enterprises and other entities was conducted through the study of the frequency, forms and reasons of contact. Although the original study incorporates contacts between sports and adventure tourism firms and public sector organisations, for the purpose of this chapter only the relationships with other private sector firms will be discussed. The segmentation of private sector enterprises follows the United Nations World Tourism Organisation classification of the Tourism Satellite Account, which defines the core components of the tourism sector: accommodation, food and beverage, transportation, tour operators and travel agencies, rent-a-car, cultural and recreational services. The fourth section concerns the establishment of partnerships and networks, namely its importance, role and advantages. The fifth section relates to the perceived role of these firms in local

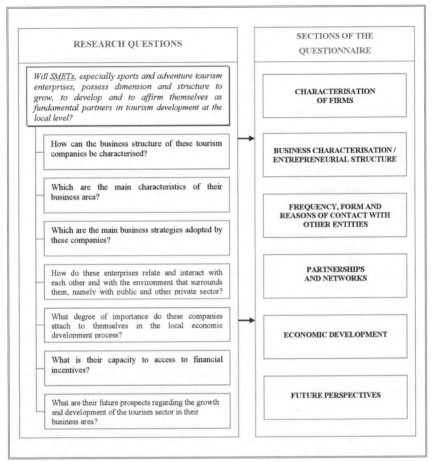

Figure 8.2 Methodology of the study

economic development, and the last section infers their future growth and development prospects.

Portuguese Tourism SMEs: The Case of Sport and Adventure Tourism Companies

Characterisation of enterprises

According to the definition of the European Commission, which considers micro enterprises the ones with fewer than ten employees and with an annual turnover and/or balance sheet not exceeding 2,000,000 euros (Commission Recommendation 2003/361/EC), the majority of the firms in this sub-sector can be considered micro enterprises. In fact, more than 90% of the respondents have a turnover of less than 2 million Euros; moreover, half of them showed a turnover less than 50, 000 euros. Accordingly, 91% of all enterprises employ fewer than ten people, but the bulk of those enterprises (nearly 80%) comprise only four employees or less.

A European Union study identified that the main business objectives of small and medium-sized firms in Europe are to grow, to increase profits, to innovate, to improve the quality of the service, to survive and to consolidate their business (Observatory of European SMEs, 2003). According to those objectives, the first priority of sports and adventure tourism firms is to grow. A large number of sports and adventure tourism enterprises is also highly innovative and want to improve the quality of their services.

Regarding their main target markets, and contrary to the tendency of small enterprises in the services sector, which are mostly oriented to local markets (Hassid, 2003), it was observed that the main target of these enterprises are the national and the international markets. This situation demonstrates the capacity of some of these companies to promote their activities abroad and consequently their ability to attract new customers, despite their small scale and family structure.

Formal and informal networking relationships

Sports and adventure tourism enterprises consider it extremely important to not 'be alone' in their business area; thus they have started gradually, on an individual basis, to associate themselves to several entities. Some of these organisations, in spite of already having some presence and importance, do not fully represent the main interests of these enterprises. However, it should be pointed out that the effort made by these companies to be affiliated with entities that can support them in some way is important. Nonetheless, over 60% of these enterprises do not belong to any tourism association at national, regional or local levels. This can be explained by the feeling of a lack of representativeness, since they consider that the existing associations do not fully protect their interests. A similar situation had already been found by Costa (1996) when a study about the Portuguese Regional Tourism Boards was conducted, and a lack of representativeness of the 'official' tourism representatives was also found.

It should nevertheless be pointed out that most of these unaffiliated enterprises only possess one to three years of activity. In some cases this means that these new companies are still dealing with several constraints, and being affiliated to an association is not a top priority. In other cases, new companies are not aware of the existence of these entities.

Those companies affiliated with tourism associations pointed out some positive aspects. For them, the main advantage is the representativeness and credibility that this relationship brings. Given their smaller size and family business structure, some main constraints relate to a lack of credibility when dealing with suppliers and customers, and the need to be represented before governmental organisations and other private sector entities. This need is even more evident, the second major advantage pointed out by sports and adventure tourism enterprises being the influence that tourism associations can have within governmental bodies. For nearly 17% this factor is extremely important given that, on their own, it would be extremely difficult to have their opinions heard or to express themselves.

Two other major advantages in being affiliated are the provision of technical support and training (14%) and the possibility of obtaining updated and detailed information about the tourism sector (13%). These two advantages seem to demonstrate the need of these small-scale companies to obtain support and information, so as to have the appropriate tools to stimulate their growth and development, as well as to define properly their management and business strategies.

The possibility of engaging in strategic partnerships was cited as a relevant reason for being affiliated in sector-based associations. In fact, the possibility of establishing formal contacts with other companies and exchanging views about problems, and also having informal contacts, allows the establishment of strategic partnerships that evolve in a natural way. Another particularly important advantage has to do with the institutional and legal support that it is granted by associations to their associates. Being small-scale companies they have limited human resources; for this reason the institutional and legal support granted by associations is extremely important.

Other advantages, such as gaining more negotiation capacity, benefiting from common services and undertaking joint promotion of their products, were also pointed out when considering a group of enterprises rather than thinking on an individual basis. Only a minor percentage of companies pointed out disadvantages in joining an association. The main disadvantages relate to the need to pay fees, and the lack of time and availability to participate in activities and meetings.

Type and characteristics of contacts between stakeholders

Another aspect considered in the study was the firms' relationships with other tourism industry partners. The frequency of contacts among them is very high, especially with other private sector companies, thus showing an interesting business cooperation (see Figure 8.3). It is interesting to note that the frequency of contacts reflects the degree of importance attached by these enterprises in establishing partnerships and/or business agreements with other entities working within the tourism sector. Accommodation services, travel agencies and tour operators, and food and beverage services are among the entities that have the highest percentage of contacts established. The intensity of contacts is also high, since these entities are contacted with more regularity. These relationships could therefore be encouraged in order to establish a formal tourism network. Although interactions with the public sector are not discussed here, it is interesting to note that contacts are not that frequent, despite the important role it plays in the tourism sector. This might be because the public sector embodies the major constraints that enterprises face, which are related to the lack of support from governmental entities, funding and legal matters.

Regarding the way that contacts are established (see Figure 8.4), it is noticeable that there are differences according to the type of organisation. Generally speaking, sports and adventure tourism enterprises contact

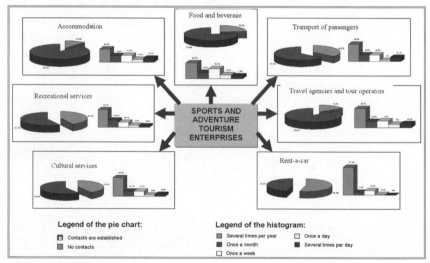

Figure 8.3 Frequency of contacts between tourism stakeholders

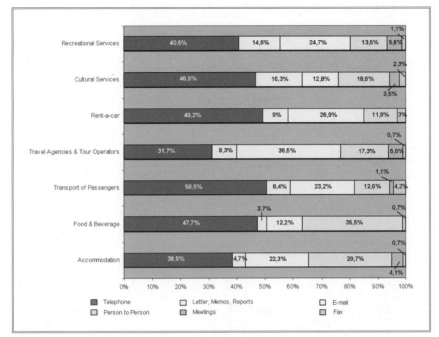

Figure 8.4 Form of contact between tourism stakeholders

other private sector entities by telephone, personally or by e-mail, while contacts with governmental organisations are made essentially through letters, memos or reports, or by phone. Naturally, the method of contact used depends mostly on what it is necessary to transmit or to ask. And, in fact, when it is necessary to contact public sector organisations it is important to officialise everything; in that sense it is convenient to use traditional methods such as letters, memos or reports.

When dealing with private sector entities, in most cases it is not so important to make an official contact. Sometimes there are personal ties among the people involved, which facilitate the form of contact. Another aspect, that has to do with the type of information requested, relates to the fact that the exchange of information needs to be fast and effective. Generally speaking, the use of telephone is extremely high in every contact made, except when contacting travel agencies and tour operators, where e-mail is the preferred method of contact (37%). With food and beverage and accommodation services, sports and adventure tourism enterprises use

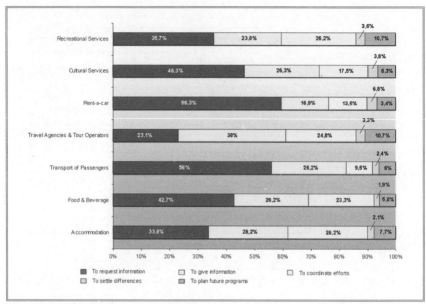

Figure 8.5 Reasons why contacts are established between tourism stakeholders

personal contacts on a higher percentage than with other services (36 and 30%, respectively).

Frequency and form of contact obviously depend on the reasons for contact. A group of reasons were defined that could explain why these enterprises get in touch with each other: to request or to give information, to coordinate efforts, to settle differences or to plan future programmes (see Figure 8.5). It can be inferred that these enterprises contact other entities mainly to request information, except the contact with travel agencies and tour operators which is established to provide information. This is one of the reasons why e-mail is one of the preferred methods of contact, as it provides evidence of the information sent. Another important reason relates to the need for coordinating efforts, especially with accommodation and food and beverage services, travel agencies and tour operators, and recreational services. This demonstrates that sports and adventure tourism enterprises have a closer relationship with these entities, which means that there is great potential for increasing contacts among several service providers and entities, since they all can benefit from this proximity.

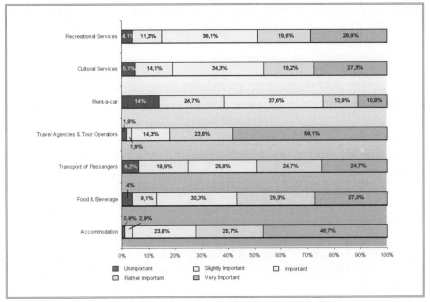

Figure 8.6 Importance of the contacts established with other organisations

Importance of the establishment of relationships with other organisations

Sports and adventure tourism enterprises consider the establishment of partnerships and/or business agreements with other entities working within the tourism sector to be important (see Figure 8.6). More specifically, 96% of them consider important, rather important or very important the development of partnerships with travel agencies and tour operators, and with accommodation services. Despite the fact that these two sub-sectors have the same level of importance, it is noticeable that their significance is attributed for different reasons. The role of travel agencies and tour operators is mainly that of intermediation between these companies and the final consumer, whereas the accommodation sub-sector represents one important service supplier.

Food and beverage services represent another important partner since nearly 87% of sports and adventure tourism enterprises consider fundamental the development of partnerships with this tourism sub-sector. Accommodation and food and beverage services, besides satisfying the basic needs of the consumers of these enterprises, comprise the elements of the static anchoring process of tourists (Breda *et al.*, 2006). Other services also integral to the core of the tourism sector but constituting the dynamic anchoring process are considered relevant partners for these small and micro enterprises, namely cultural and recreational services. Although transportation services and rent-a-car companies are considered important, they are the least-mentioned items.

Importance of the existence of a tourism network

Regarding the establishment of tourism networks, 92% of the enterprises attach a great importance to it; moreover, 54% consider it very important. This situation clearly evidences the need to develop a tourism network in their business area, although these enterprises believe it is also necessary to create an organisation that could represent their interests and promote the creation and development of partnerships between all tourism sub-sectors.

Almost all sports and adventure tourism enterprises (99%) consider that one of the main advantages of being included in a network of organisations is the capacity to influence tourism policies at the local level (see Figure 8.7). Nearly 95% of respondents believe that networks stimulate better-informed solutions, which are brought into the decision-making and decision-taking processes. Moreover, 88% believe that the establishment of tourism networks is important to enable better geographical management and distribution of resources, which in turn will influence the efficiency and effectiveness of tourism policies. In this sense, 94% of the enterprises consider that improving their efficiency and effectiveness is one of the main advantages of the development of tourism networks.

Results also show that 91% of the SMEs attach a great importance to networks as a means to increasing competitiveness in the travel and tourism market, which will improve their innovation capacity (as mentioned by 91% of respondents). Indeed, the literature suggests that networks and clusters can help in the innovation process of tourism SMEs (see Ahuja, 2000; Fukugawa, 2006; Haga, 2005; Holbrook & Wolfe, 2005; Liyanage, 1995; Matteo *et al.*, 2005; Perryman & Combs, 2005). Novelli *et al.* (2006) demonstrate that there is a correlation between competitiveness produced by clusters and the ability of members to improve their services and products through inter-firm linkages and innovative business approaches.

Nearly 88% also consider the development of a tourism network to facilitate the internationalisation of enterprises to be important. It can be inferred that the existence of a network is helpful for small firms in reducing the risk associated with internationalisation. This result is in accordance with many studies which show that one way to acquire internationalisation-related competences is through the promotion of networking, in the form of formal

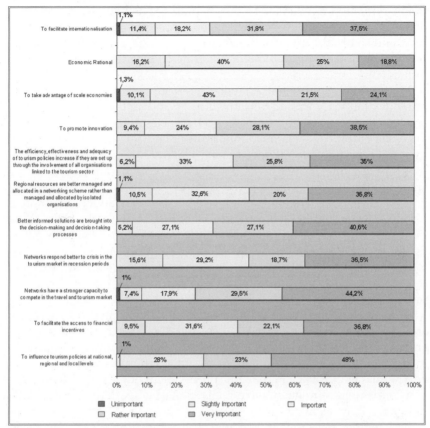

Figure 8.7 Advantages of being included in a network of organisations

and informal links (see Chetty *et al.*, 2003; Coviello & Munro, 1995, 1997; Etemad, 2004; Etamad & Wright, 2003; Forsgren *et al.*, 2005; Hassid, 2003; Johanson & Vahlne, 1990; Lindstrand, 2003; Madsen & Servais, 1997; Sharma & Johnson, 1987). Knowledge is a key resource in the internationalisation process of firms and can be attained through relationships between firms in a network.

Conclusions

Innovation is increasingly recognised as a major catalyst for productivity and regional growth. This paper has highlighted the fact that successful tourism networking requires a high level of cohesion, innovation and knowledge. As the chapters in the book show, research on the importance of networks in the tourism sector has grown considerably, and so has the interest of its benefits to SMEs, which dominate the sector. The process of innovation illustrates the benefits of networks to induce competitiveness.

Sports and adventure tourism enterprises support the theoretical view of innovation networks, underpinned by the advantages of information and knowledge exchange, productivity efficiencies, and competitive advantage. The empirical research demonstrates that SMEs clearly assume an important role in economic development, coordinating much of their business through informal networks. Sport and adventure tourism enterprises recognise the importance of networking and establishing formal contacts, which facilitates knowledge exchange, joint promotion and distribution of resources. Also highlighted is the ability of formal networks to influence local tourism policies and to facilitate the internationalisation of enterprises. To survive in an increasingly competitive and global environment, tourism enterprises, particularly small and micro enterprises, increase their productivity and strive in the regional and international markets through networking.

Restructuring and cooperation mechanisms help enterprises to grow and to adapt to changes. Due to the recognised interest for business cooperation, more effective and efficient relationships emerge as a means to increase innovation and competitiveness. Innovation in tourism is seen to be a permanent, global and dynamic process. This empirical study reveals that tourism entrepreneurs have realised that networking innovation is becoming a key element to survive and compete in a dynamic and radically changing environment. Theoretical and empirical research is needed in the construction, evolution and benefits of innovative networks tourism for SMEs.

Chapter 9

International Tourism Trade Networks: The Case of the Chinese Inbound Travel Trade to Australia

GRACE WEN PAN

International Tourism Trade Networks

Business networking in the tourism industry is a fairly new area of academic interest and research. Crotts *et al.* (2000) see the main purpose of forming networks as making the firms involved in the network become more competitive. The advantage of forming networks in the tourism industry is that firms involved in the relationship contribute complementary components to achieve a level of satisfaction for all firms involved.

However, 'business structures and relationships are influenced by many social and economic factors, but are significantly influenced by culture, particularly values' (Fulop & Richards, 2002: 274). How different players in various settings interact with each other, and how this is affected by social and cultural differences, has been acknowledged as important, especially when establishing new business relationships in an international context (Björkman & Kock, 1995; Holmlund & Kock, 1998). Although there is considerable research into business networks in the international context, little research has been conducted into how business networks are managed in the tourism and hospitality industry.

Cultural Studies in Tourism

Tourism is primarily a socio-cultural activity, and socio-cultural needs and psychological experiences are thus more valuable to tourists than material needs (Reisinger & Turner, 1998). Approaching a new market without considering the cross-cultural implications of the global nature of the travel industry may lead to misleading results or an unsuccessful marketing campaign (Money & Crotts 2000). Most of the academic literature has reviewed and discussed culture as an elementary factor impacting on international tourist markets (for instance, King & Choi, 1999; Pearce *et al.*, 1998; Reisinger & Turner, 1997). However, these studies have concentrated more on the impact on international tourists and the consequent marketing strategies. Limited research has been conducted with regard to the role of culture in the international tourism trade, although March (1997) states that the role of culture in the development of international tourism markets should be considered when exploring the influence of business culture on developing and maintaining business-to-business relationships.

A case study approach was adopted in this study to investigate the role of networking in the Chinese inbound travel trade to Australia. In particular, this study examines the reality of network development in the cross-cultural context, and fills the gap by specifically investigating how cultural differences impact on the development of partnering relationships between the two key players – Australian inbound tour operators and Chinese travel agents – in the Chinese inbound travel trade to Australia.

The Case of the Chinese Inbound Travel Trade to Australia

The growth of China's economy is leading to a rapid increase in international travel from a burgeoning middle class. The Chinese Government did not loosen the restrictions on the outbound tourism market until 1983, when a 'new' form of travelling under the Approved Destination Status (ADS) scheme was introduced. ADS is an administrative measure by means of which the Chinese Government permits its residents to travel to selected countries for personal and leisure purposes, usually on all-inclusive package tours. Eighty one countries have been granted ADS. China has been recognised as one of two major emerging outbound tourism markets in the world. The World Tourism Organisation (2003) has forecast that China will have 100 million outbound travellers, and become the fourth largest source of outbound travel in the world by 2020 (see Figure 9.1).

Australia was the first western country opened up to the Chinese outbound tourism market, followed soon afterwards by New Zealand in 1999. Mainland China has been acknowledged as an important emerging market by the Australian inbound tourism industry. The number of Chinese arrivals in Australia has increased at an average rate of 25.9% each year since 1985, reaching 285,800 in 2005 (Australian Bureau of Statistics, 2005) (See Figure 9.2). Among these short-term Chinese visitors to Australia, over 160,000 Chinese ADS visitors in approximately 10,500 groups have travelled to Australia since 1999 (Department of Industry,

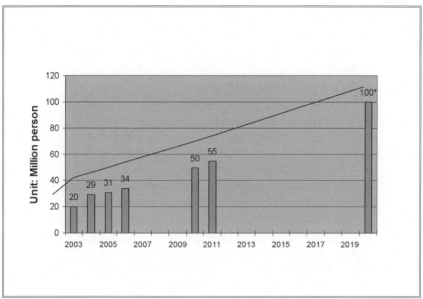

Figure 9.1 Anticipated increasing number of Chinese outbound travellers from PRC. (Note: * forecast figure)
Source: China National Travel Association (2005), World Tourism Association (2003)

Tourism and Resources, 2005). While Chinese visitors accounted for only 1% of all international tourists visiting Australia in 1995 this figure is expected to grow at 16% per annum for the next decade, reaching approximately 1.2 million Chinese visitors per year by 2014 (Tourism Forecasting Council, 2005). Thus, Australian tourism operators and State Tourism Offices see the Chinese outbound tourist market as a great opportunity to tap into, while the numbers of most of the Asian tourism markets to Australia are somewhat stagnant. Nevertheless, a number of problems in the operation of ADS within Australia have become apparent and are now being addressed (Department of Industry, Tourism and Resources, 2005).

Although Australia is geographically located in the Asia-Pacific region, it is a Western society and its business executives in the export sector confront the same problems as do those from the USA and Europe, and share similar attitudes to management. Tixier (2000) undertook research into Australian management efforts to internationalise business. The investigations revealed that levels of international knowledge in Australia are biased in favour of Europe because Australia is a country with an

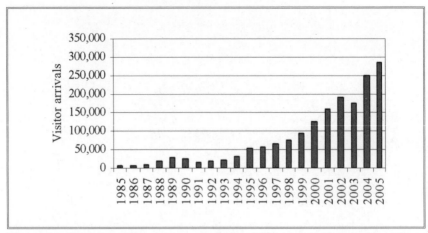

Figure 9.2 Short-term movement – arrivals of Chinese visitors to Australia 1985–2005
Source for data: Australia Bureau of Statistics (1989; 2005)

Anglo-Celtic heritage. This is consistent with the nation's poor record in developing significant and meaningful changes in attitudes towards, and knowledge and awareness of, Asia. Many Australians fail to take into account the cultural context in which Asian business operates and, as a result, they encounter negative Asian business experiences, based on a lack of understanding of Asian business psychology, which is as diverse as there are cultures and sub-cultures (Tixier, 2000). It appears that this is damaging many firms' ability to operate successfully in the region.

Business networks in the Chinese inbound travel trade to Australia

While Hall (1995) has identified the characteristics of the Australian tourism market system, there has been little research into the Chinese tourism market system. This case study focuses mainly on the partnering relationship between Australian inbound tour operators and Chinese travel agents, with particular reference to the cross-cultural context. Figure 9.3 provides a relationship map of key stakeholders in the Chinese inbound tourism trade to Australia. Australian inbound tour operators, as suppliers, deal directly with the Chinese authorised travel agents, as buyers, in the Chinese inbound travel trade to Australia. Australian tourism operatives, such as attractions, airlines, duty-free shops, accommodation, restaurants and bus companies, are highly dependent on Australian inbound tour operators to obtain access to the Chinese tourism market.

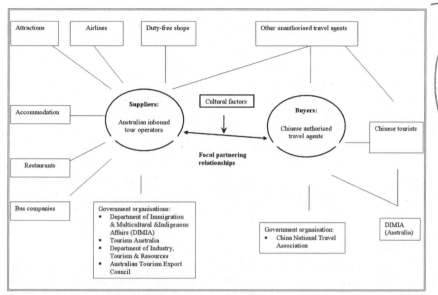

Figure 9.3 The dynamics of business network relationships in the Sino-Australian international travel trade
Source: Pan *et al.* (2007)

Australian Government organisations, such as the Department of Immigration and Multicultural and Indigenous Affairs, the Department of Industry, Tourism and Resources and Tourism Australian, interpret government policy in order to provide inbound travel-related policies and assist in promoting Australian inbound tourism markets. On the Chinese side, Chinese tourists mainly rely on Chinese authorised travel agents to organise tours for them to travel overseas. Australian operators may deal directly with unauthorised Chinese travel agents to gain niche business other than ADS groups, such as technical visits, incentive tours and study tours. However, even if unauthorised Chinese travel agents are successful in obtaining the business of Chinese leisure tourists who intend to travel to Australia with ADS visas, these unauthorised agents still have to pass their business to authorised travel agents, and obtain commissions from them. Therefore, these unauthorised agents act as retailers in the Chinese inbound travel business to Australia.

In China, the Government's China National Tourism Administration (CNTA) controls and regulates the operations of outbound travel, thus Chinese tourists mainly rely on Chinese authorised travel agents to organise overseas tours for them. Australian operators may directly deal with unauthorised Chinese travel agents to gain niche business other than ADS groups, such as technical visits, incentive tours and study tours. However, even if unauthorised Chinese travel agents are successful in obtaining the business of Chinese leisure tourists who intend to travel to Australia with ADS visas, these unauthorised agents still have to pass their business to authorised travel agents, and obtain commissions from them. Therefore, these unauthorised agents act as retailers in the Chinese inbound travel business to Australia. Nevertheless, the bulk of Chinese inbound travel business is operated between Chinese authorised travel agents and Australian nominated inbound tour operators. This case study focuses on the partner relationships between authorised Chinese travel agents and nominated Australian inbound tour operators, the two crucial players in the Chinese inbound tourism trade to Australia.

Method

This study adopted a qualitative approach (in-depth interviews) to investigate Chinese inbound travel business to Australia by interviewing Australian nominated inbound tour operators and Chinese authorised travel agents. As this study was cross-cultural and had an emphasis on Chinese respondents, the interview approach was considered the most appropriate within the constraints of the limited opportunities in the field (Malhotra *et al.*, 1996; Pyatt 1995). Furthermore, it should be noted that the interview issues included a discussion on *guanxi*, which has been acknowledged as a sensitive issue in discussions in China. Both Guthrie (1998) and Bian (1994) suggest using an in-depth interview format to obtain more valuable and meaningful information on *guanxi*. Therefore, in-depth personal interviews were deemed the most appropriate approach to employ in order to investigate our research question.

Of the nominated operators provided by the Australian Tourism Export Council (ATEC), only 30 dealt with the Chinese inbound tourism market at the time of data collection in 2002. In China, CNTA had authorised 20 travel agents to handle this business. Two lists were used to contact potential interview cases. Eleven Australian inbound tour operator interview cases and eleven Chinese travel agent interview cases were used in this study.

Standardised open-ended questions were developed in English through consultations with industry (both in Australia and China) as well as drawing on the literature on business relationship development. An interview instrument containing standardized open-ended questions was used in this study. Selected sample interview questions are provided in Table 9.1. Managers from authorised Chinese travel agents were interviewed in

Table 9.1 Key aspects studied and selected questions examining the Chinese inbound travel trade to Australia

Key aspects studied	Questions
Current situation	Background information on company
Network relationships	The process of developing network relationships
Guanxi	The role of *guanxi* in the process of developing network relationships
Cross-cultural factors	Differences and difficulties in doing businesses
General questions	Supplementary part of interview to establish what managers consider important in developing business network relationships

Note: * All questions were translated into Mandarin and presented to interviewees.

Mandarin rather than in English, although most of them had a basic understanding of English. Pre-testing of the interview questions for 'linguistic equivalence' (Maholtra *et al.*, 1996: 25) was undertaken. The interview instrument was translated into Mandarin by the author, who is a native Mandarin speaker, and then back-translated to English by an accredited Chinese–English translator to ensure accuracy and equivalence. Content analysis was adopted in this research to analyse the data and was undertaken with the aid of NVivo2.0 software for data management purposes. The findings of the study demonstrate that the process of developing partner relationships is highly culturally 'contaminated'.

Profile of businesses involved in the partnering relationships

The Australian inbound tour operators' principal role was to coordinate with other tourist product suppliers, such as hotels, restaurants and coaches, to provide land services to international tourists. The Australian inbound operators dealt with various inbound markets (not just China). Consistent with what Lambert (1996) claims – that in Australia, 80% of tourism operators are privately owned enterprises with micro-business, and some are small companies with fewer than 20 people – all the interviewed Australian inbound tour operators were Small and Medium Enterprises (SMEs), a few of which were family operated businesses, with the range of business mainly focused on inbound travel operations. Furthermore, all Australian operators were of Chinese descent, five of whom were from Mainland China.

Ten out of eleven interviewed Chinese travel agents were State Owned Enterprises (SOEs) (except Case CA) with a portfolio of businesses including both inbound and outbound travel. Australian inbound tour operators had a large number of counterparts, while Chinese travel agents had a limited number of Australian counterparts, ranging from one to five. The Chinese travel agents, as buyers, had more bargaining power, whereas the Australian inbound tour operators, as suppliers/sellers, needed to find as many clients/customers as possible, both in the Chinese market and elsewhere.

The Role of Cultural Differences on the Development of Chinese Inbound Travel Trade to Australia

The large-sized Chinese travel agents and small-scale Australian inbound tour operators form a contrast in terms of size. Although some Chinese travel agents stated their preference for having similar-sized Australian counterparts, the fact that most Australian inbound tour operators are SMEs dictates that such preferences may become 'mission impossible'. Therefore, this study concurs with the literature that size is not a critical issue for competitiveness in networking relationships (Howard, 1990). In fact, the real moderators, as independent factors externally impacting on the whole process of developing partnering relationships between Chinese and Australian counterparts in the Chinese inbound tourism market to Australia, are *ethnic preferences* and *regional cultural differences*, which are discussed below.

Ethnic preference

Previous literature has referred to cultural distance, describing it as '... the degree to which the norms and values of the two companies differ because of their place of origin' (Ford *et al.*, 1998: 30). One of the major barriers to successful performance in cross-national business relationships is the degree of cultural distance between two counterparts (Ford *et al.*, 1998; Williams *et al.*, 1998). Cultural distance exists between western nations and China, and can cause cultural shock, such as when western expatriates work in China (Hutchings, 2002; Kaye & Taylor, 1997). Despite the language differences, cultural distance was evident in this study between Australian Caucasian operators and Chinese travel agents, and this distance impacted on their ways of doing business. Some Chinese travel agents used Caucasian partners; however, their experience showed that Australian Caucasian operators have different mindsets from Chinese travel agents, as sometimes Chinese travel agents' requests were considered unethical by their Australian partners, but considered only as a normal request or a favour by the Chinese travel agents. The respondent CK[1] told the following story:

> We used to have a Caucasian operator in Melbourne. He is a very nice man, but he refused to do business the way we suggested. Therefore, our business relationship was terminated in the end. For example, we asked them to include the amount of tips in the total receipt, and he refused to do this. Our clients like to have some pocket money, and they would like to include the pocket money in the package amount – he refused again. He thought that it was a ridiculous way of doing things. However, he has to understand that some people in China travel using government or company's money, and they do not want to pay anything. Hence, as you can see, we (Chinese and Caucasians) have different mindsets.

This story indicates that it is important for Australian operators to understand the Chinese cultural background and the nature of organisational structures in China. Different ethical beliefs were the major reasons for Chinese travel agents preferring to use Australian Chinese operators with the same or similar ethnic background, even though there were still some differences as previously discussed. These preferences effectively exclude Caucasian operators from such partnering relationships.

Another reason Chinese travel agents prefer to have Australian Chinese as their partners is because of the perceived cultural closeness. In the international business setting, cultural closeness is the consequence of the reduction of cultural distance between two counterparts who are in two different countries (Adler & Graham, 1989; Swift, 1999). Having cultural closeness made the Chinese agents feel confident in doing business with Australian Chinese operators. Even the Australian operators noticed that the Chinese travel agents like to do business with Australian operators from the same cultural background, and Respondent AR explained, '...it is easier to communicate in the same language, and we share the same culture so that we can easily exchange our thoughts'. Therefore, ethnic closeness is one of the key criteria used by Chinese travel agents to choose their Australian operators. In particular, Australian operators originally from China have an advantage when seeking partnerships with Chinese travel agents.

Even though all the Australian operators in this study are first generation Australian Chinese migrants to Australia, only some of them are originally from mainland China, while others are from Taiwan and Hong Kong, and one is from Brunei. They are aware that the ethnic Chinese community is not homogeneous and they differentiate themselves according to their origins, whether in China or outside China (East Asia

Analytical Unit, 1995). The interview responses from Chinese travel agents reflect that Australian Chinese operators from Taiwan and Hong Kong think and approach the mainland Chinese differently when establishing business relationships. Respondent CG provided an example, elaborating on the distinctive way of doing business of Australian operators who are originally from Taiwan, stating:

> We have totally different ways of doing business (from these operators originally from Taiwan). We usually discuss the business in the daytime, and then perhaps have dinner. That is about it. However, those Taiwanese operators came to meet us around lunch time, then afternoon tea, then dinner, then dancing, whereas we are not in the habit of having afternoon tea. Ok, we will go with them, but we are really not used to the way they do things. From our point of view, business is business, and we separate it from social entertainment. I think that we are very different from Taiwanese in doing business on this matter.

Respondent CG further commented, 'We cannot use the way of doing business in Taiwan to apply to our way of doing business'. It seems that those Taiwanese who live in Australia still have the traditional mindset about the inefficiency of working in China and the importance of gaining *guanxi* through social activities. Therefore, these different ways of doing business may hinder the formation of business partnering relations between Chinese and Australian counterparts from Taiwanese descent.

In addition, due to different communication styles, it appears that Chinese travel agents prefer to have Australian operators who are originally from mainland China rather than from Hong Kong or Taiwan. Previous literature, such as Hall (1976) and Harris and Moran (2000), identifies communication differences between people in high context cultures where people are deeply involved with each other, and low context cultures where people are highly individualised, even alienating and fragmented, but not within one single context culture. However, this case study reveals that there are significant communication difficulties between Chinese travel agents and Australian operators, despite all of them being able to speak Mandarin and belonging to high context communication cultures. These communication differences include different wording, expressions and different ways of talking between mainland Chinese and Chinese from Hong Kong and Taiwan. Both Respondents CE and CG noticed different wording and expressions used by mainland Chinese compared with Chinese from Hong Kong and Taiwan. Respondent CG remarked:

Although all of us (including mainland Chinese and Chinese from Hong Kong and Taiwan) speak Mandarin, they (Chinese from Hong Kong and Taiwan) use different wording and expressions, and there are differences in the ways people talk between people in mainland China and people in Hong Kong and Taiwan. Therefore, we have some communication barriers in contacting each other. Although people from Taiwan and Hong Kong speak Mandarin, I do not understand them well.

These differences directly impact on the Chinese travel agents' decisions in choosing their Australian counterparts. It appears that the communication problems between Chinese and Australian partners extend beyond the language problem. The Chinese travel agents prefer to deal with Australian operators who are originally from mainland China simply because of the more comfortable communication experience. Hence, co-ethnic Chinese background becomes a preference for most of the Chinese travel agents. It is apparent that having someone who is able to understand the culture and speak the language is a facilitating factor. It smoothes the negotiation of key points and differences between partners. Thus, Australian inbound tour operators should consider engaging a business development manager with such skills.

Regional cultural differences

Regional cultural differences within China draw the attention of Australian operators to subtle variations in business practices in China. Little research has been conducted on regional cultural differences within China, with the exception of Selmer (1997), who addresses the different regional cultural stereotypes among Chinese. For example, there is a reported antipathy between Cantonese and Shanghainese (Selmer, 1997). This case study confirms that Chinese travel agents do not constitute a homogeneous group, and there are regional cultural differences within China. Two different stereotypical ways of doing business in Beijing and Shanghai emerged from this study. People in Shanghai are more efficient and less bureaucratic compared with those in Beijing. Respondent CG provided an example explaining the differences between doing business in Shanghai and Beijing, stating:

In Shanghai, if we are going to sign a contract, each party will have a copy of the draft of the contract prior to the meeting. On the next day, if everyone agrees with everything, we can sign the contract. Once the contract is signed, people in Shanghai follow the contract clauses.

However, it is a different story in Beijing. First of all, the list is not dispatched before the meeting. The meeting becomes very bureaucratic, and discussions are carried on and on. Even after the contract is signed, some terms in the contract can still be altered.

The study shows that Chinese travel agents also have their regional preferences, and prefer to form partnering relationships with Australian operators based on dialects. That is, Chinese agents in Shanghai tend to find Australian partners who are originally from the same area, as do agents in Beijing. Chinese travel agents prefer to have Australian counterparts who were originally from the same region. Respondent CK, who was based in Shanghai, responded, 'We would prefer to have Australian operators from the same region (Shanghai). One of our current Australian partners is originally from Shanghai. One of the major reasons we use that operator is because he is Shanghainese, and we can speak in Shanghai dialect. He also knows the ways of doing things here in Shanghai very well.'.

However, it seems that most Australian operators still have not realised that the Chinese culture is embedded in different regional cultures, and they still perceive Chinese travel agents as a homogeneous group without considering regional cultural differences. A more heterogeneous approach makes more sense when Australian operators try to establish and develop relationships with Chinese travel agents. Alternatively, employing people from specific regions or who speak specific dialects seems crucial to overcoming this significant impediment to developing business relations.

Guanxi

The Chinese business network has been broadly accepted as a *guanxi*-based business network (Chen, 1995; Pye, 1985; Tung, 1991; Wong & Tam, 2000). Person-to-person relationships are called *guanxi* in Chinese society. Although the concept of *guanxi* has been used for more than 2500 years, it did not become the focus of attention until the Cultural Revolution. Yang (1986), in addressing the reason for the popularity of *guanxi* in China, suggests that the breakdown of social order during the Cultural Revolution forced people to rely on *guanxi* (connections) rather than the state organisational structures to secure their everyday survival. For example, during the Cultural Revolution, people faced the prospect of sending their children to some very remote areas of the countryside, hence whether they had *guanxi* with the central decision-making person or not played a critical role in deciding the future of their children. Those people who did not have *guanxi* tried to obtain it by exchanging gifts and favours among personal connections to find ways to deal with the crisis.

Guanxi is an intricate and illusive concept (Dunfee & Warren, 2001). Tsui and Farh (1997) comment that there is no consensus in the translation or definition of the term '*guanxi*'. *Guanxi* can be defined from both a macro and a micro perspective. From the macro perspective, the meaning of *guanxi* refers to the existence of some kind of relationship between individuals or individuals with objects; it can be referred to as any kind of relationship (Chinese Contemporary Dictionary, 1983: 407). As long as it is called relationship in English, it can be translated into *guanxi* in Chinese.

It is acknowledged that doing business in China is complicated, and that the Chinese business network is a *guanxi*-based network. Studies of Chinese networks have mainly focused on overseas Chinese networks (family networks), friendship and *guanxi* to gain an understanding of Chinese businesses (for instance, Blackman, 2000; East Asia Analytical Unit, 1995; Haley & Tan, 1999; Hutchings, 2002; Redding, 1993).

The changing role of *guanxi*

Since the implementation of the open door policy, China has been transformed from its so-called self-sufficient and self-contained economic development model to a market-oriented economy model. China has made great economic progress, becoming a major player in world trade in the last two decades, and significant changes have taken place in the domestic business environment. Dunfee and Warren (2001) list some of these changes as: increased privatisation of business firms; a movement towards more emphasis on the rule of law; changes in forms of business operation and corporate governance; increasing competition among business firms; and increased foreign investment. Westerners need to adjust their previous perceptions to the current situation in China.

The content of *guanxi* practice is changing with China's transition from a command economy to a market economy. There is a move from exchanging gifts, or doing favours (Yang, 1994), to actions more focused on the development of business relationships (Guthrie, 1998). The study conducted by Guthrie (1998) ends up with two different responses from managers in China. One group views *guanxi* as an important aspect of market economies; they also state that personal relationships enhance business and can serve as an advantage in the increasingly competitive markets during the economy transition period. The other group views *guanxi* as decreasing in importance in China, while price, quality and service are the primary factors which shape market relationships and play an increasing role in the economic transition. Hence, the researcher feels it important to discuss *guanxi* separately, as the role of *guanxi* and its patronage is changing in modern China (Pan *et al.*, 2007). This situation needs to be drawn to the attention of Australian inbound tour operators and marketers in particular.

In this case study, both Australian and Chinese counterparts supported the view that *guanxi* goes beyond personal relationships and includes both personal and business relationships, even though most of the literature (for instance, Bian, 1994; Davies *et al.*, 1995; Leung *et al.*, 1996) supports the meaning of *guanxi* as personal relationships. Furthermore, the findings from the Chinese travel agents reveal that they have mixed responses with respect to the involvement of personal relationships in the meaning of *guanxi*, and the importance of *guanxi* in the process of developing part-nering relationships with their Australian counterparts. Although a few still stated that it was difficult to separate business relationships from per-sonal relationships, some of the Chinese travel agents clearly drew a boundary between personal relationships and business relationships, whereby corporate behaviour took precedence and personal relation-ships were less important in the process of developing partnering rela-tionships. For example, Respondent CK commented, 'Forming business partnerships is a corporate behaviour, and there should be no personal relationships involved. We do not talk at a personal level.' Respondent CI further explained the reason why they considered *guanxi* as not a part of corporate behaviour, saying:

> As our company is a state-owned enterprise, our behaviour is corpo-rate behaviour, and our decision in choosing a partner company has to consider quality of services and company profit prior to considering *guanxi* (personal relationship). Our agent is not like some agents who are contracted by a few people. These companies are like private companies where *guanxi* (personal relationships) plays an important role.

Hence, it seems that *guanxi*, at least publicly, does not play an important role in those large-sized state-owned travel agents where price and profit are considered most important. This finding is consistent with Guthrie's (1998) recent interviews in China, where he found that in market relation-ships, the importance of *guanxi* is secondary to market imperatives of price and quality. This finding further reinforces the evidence that Chinese travel agents are 'Westernising' their ways of doing business by priori-tising corporate relationships over personal relationships.

Nevertheless, in some situations *guanxi* still plays a role in facilitating business relationships between Chinese and Australian counterparts. This study identifies that having *guanxi* helps consolidate the partnering relationship and adds value to the relationship. Respondent CH remarked, 'The personal relationship plays a catalytic role in the process of establishing a business relationship with our Australian counterparts.' This finding is a subtler refinement of the *guanxi* concept and an important form of knowledge for those entering this field of business. For example, when the price and quality of services offered by the Australian operators were the same or similar, Chinese travel agents would establish business partnering relationships with those Australian operators with whom they had a *guanxi* relationship. This situation reiterates one of the benefits of having *guanxi* relationships, and demonstrates how they add value to the business relationship (Yau *et al.*, 2000). These Chinese agents would therefore establish their business relationships with the operators they had *guanxi* relationships with. Respondent CB emphasised, 'Of course, if we know the operators well, it will make it easier for us to cooperate in all aspects and will make the partnership smooth and happy.' Hence, these findings further demonstrate that China is becoming a more complex and fragmented society under the influence of western business and markets.

Conclusion

Developing partnering relationships cross-nationally in the tourism trade industry is a complicated process, particularly when it involves two counterparts from totally different cultural backgrounds. This case study demonstrates the complexity and difficulties in the process of developing such network relationships in the tourism industry. It can be concluded that the process is embedded with cultural factors, such as *guanxi*, *ethnic preferences* and *regional cultural differences*. Moreover, the culturally embedded nature of partnering relationships between Chinese and Australian counterparts does not mean that every single step advancing this business relationship is embodied with cultural characteristics. In fact, some of the features do not have cultural aspects at all. For example, with the economic transition in progress, markets in China are becoming increasingly competitive, focusing primarily on quality of services and the pricing issue rather than *guanxi*.

The identification of Australian Chinese operators' acculturation to the Australian culture further highlights the major impediments to establishing and developing international business network relationships between Chinese and Australian counterparts: different ways of doing business, communication problems and misunderstanding of the role of *guanxi* and its patronage in developing the partnering relationships. Hence, education and training programmes need to be provided to educate Australian operators and Chinese travel agents regarding different business ethics and communication differences. It is evident from this case that the cultural differences extend beyond language differences. Understanding Mandarin is just a first step in approaching Chinese counterparts. More importantly, it appears that one of the most effective ways of breaking through these barriers and successfully establishing business relationships with Chinese travel agents is to build cultural affinity and to have cultural closeness and empathy. It seems that it is important to educate Australian operators, whether of Chinese descent or Caucasian, about the changing culture of China and its transition towards a market economy and the consequent effect on Chinese traditional business practices.

With the development of the Chinese outbound tourism industry, successfully developing partnering relationships will be crucial to sustain the number of Chinese tourists visiting Australia. This case study further accentuates the importance of understanding the key players in the distribution channels within the international tourism trade networks, and their different ways of doing business.

Note

1. Throughout this chapter respondents are given a code whereby the first letter indicates Australia (A) or Chinese (C) and the second letter represents an interviewee.

Chapter 10

Power, Destination Branding and the Implications of a Social Network Perspective

GUISEPPE MARZANO

The fragmented, diverse, unfocused, self-seeking and disorganised nature of the tourism industry allows the tourism destination to emerge as a priv- ileged context in which to understand how stakeholders work together while carrying different and often conflicting interests (Chamberlain 1992; Shaw & Williams, 2002). While stakeholders compete for 'bums on beds', collaboration has progressively emerged as the structure that governs the way stakeholders organise and make sense of their interdependency. Destination branding provides a context to study how tourism stake- holders relate with each other in a problem domain. Although destination branding has been widely described as a collaborative effort amongst stakeholders (Blain, 2001; Deslandes, 2003; Im, 2003; Kaplanidou & Vogt, 2003; Morgan et al., 2002, 2003; Morrison & Anderson, 2002), power is a critical matter within a destination branding process because of two different but related issues; the inorganic images expressed by the destina- tion brand influence tourists by shaping their preferences and, as a conse- quence, destination stakeholders exert power within the process of destination branding in order to be able to influence how the destination brand comes about.

This paper proposes the use of a social network perspective (Brass et al., 1998; Doreian & Stokman, 1997; Galaskiewicz & Wasserman, 1993; Kilduff & Tsai, 2003) as an appropriate approach to the study of power in destina- tion branding, both as reflected in the process of production of the brand as well as in its outcome, the destination brand. It is proposed that inclu- sion in the process of destination branding should not be considered only from a normative perspective. Cohesiveness around the destination brand has been related to the creation of brand equity (Konecnik & Gartner, 2007) and therefore it is proposed that network power in destination branding be defined as the ability to create a destination brand that reflects the values and the agenda of the largest possible number of tourism destination stakeholders. Moreover, this chapter provides a sociological perspective on destination branding and identifies power as the central issue for the understanding of the role of branding in tourism. In particular, drawing on a description of tourism as 'an industry which uses the community as a resource, sells it as a product, and, in the process, affects the lives of everyone' (Murphy, 1985: 165), power is also viewed from the perspective of interdependency and collective action, and social network analysis allows an explanation of power as inherently relational (Hanneman & Riddle, 2005)

Destination Branding

As a consequence of the impact that globalisation has on society and therefore on tourism as a social and economic phenomenon, many tourism destinations and tourism products have exhibited increased homogenisation and commodification (Pike, 2004). In this context, it has been observed that destinations mainly compete based on their perceived images relative to competitors in the marketplace (Baloglu & Mangaloglu, 2001). The strategic implication of the lack of perceived differentiation among tourism destinations, and the need to leverage the impact that inorganic images have in the tourism production/marketing/consumption cycle, increases the relevance of destination branding as a critical activity in the establishment and shaping of the image and personality of the tourism destination. Branding has been thoroughly studied in product marketing theory (Aaker, 1991, 2004, 1997; Keller, 1993b, 2003). The body of knowledge related to destination branding is still emerging and, consistent with the epistemology of tourism (Jafari & Ritchie, 1981), the understanding of the theoretical underpinnings of destination branding can be analysed from different perspectives. This chapter discusses the destination brand under a Foucauldian perspective and links this discussion with the theory of decision-making in order to provide an understanding of the effect of power on how and why a particular destination brand is created.

Destination branding is defined as the 'process used to develop a unique identity and personality that is different from all competitive destinations' (Morrison & Anderson, 2002: 17). The outcome of this process – the destination brand – is defined as a name, symbol, logo, word mark or other graphic that both identifies and differentiates the destination and has the purpose of conveying 'the promise of a memorable travel experience that is uniquely associated with the destination; it also serves to consolidate and reinforce the recollection of pleasurable memories of the destination experience' (Ritchie & Ritchie, 1998: 17). The process of destination branding can be considered as a component of a strategic tourism marketing planning process. Tourism is, however, a specialized context for the application of marketing theories and principles. It has been observed that principles of product (Calantone & Mazanec, 1991; Fesenmaier & Uysal, 1993; Middleton, 1989; Middleton & Clarke, 2001) and service marketing (Costa & Eccles, 1996; Jayawardena & Ramajeesingh, 2003; Laws, 1995; Teare, 1992) are relevant and widely applied in tourism. Tourism marketing nonetheless is 'unique and worthy of individual academic attention' (Fyall & Garrod, 2005: 37) due to the particular nature of the tourism destination. The tourism destination, therefore, is here defined as a complex entity in which tourism products and a community co-exist within the same geographically diffuse, politically bounded region but image-wise flexible territory. In this context, successful tourism marketing depends upon the quality and the structure of interrelationships, interdependency and interactions among tourism and non-tourism stakeholders. While traditional product and service marketing works on the underlying assumption of the ownership and control of the marketing strategy by the supplier, tourism marketing – and therefore destination marketing organisation – has to face the challenge of mediating between the common good and private interests. As Fyall and Garrod observed:

> the dependency upon public goods like beaches and areas of natural beauty for the success of many tourism organisations, such as tour operators, transport companies and accommodation provides, is such that any one component part of the system is dependent on the others for the system as a whole to work. (Fyall & Garrod, 2005: 37)

In this context, the destination brand is one of the components of the common good that the tourism destination as a whole represents.

The asymmetry in the benefits that different stakeholders are able to obtain from the tourism destination and from the destination brand, together with the related problems of opportunism and free riding (Ostrom, 1990), made it necessary for tourism destinations to generate a governance mechanism, able to work as a device making individual stakeholders act in the common interest (Olson, 1965: 7). Along the same line, Olson (1965) also observed that 'when a number of individuals have a common or collective interest [...] individual unorganised action [will] not be able to advance that interest adequately'. Collaboration therefore provides a governance structure (Phillips *et al.*, 2000; Wood & Gray, 1991) through which stakeholders can achieve mutually supportive pursuit of individual and collective benefits (Cropper, 1996). Nonetheless, although it was argued that 'each organisation, through the collaboration, is able to achieve its own objectives better that it could alone'(Huxham, 1993: 603), the benefits that derive from a collaborative process are not necessarily distributed equitably among the parties (Ring & van de Ven, 1994). As a consequence, the transmutational purpose inherent in a collaborative process implies that stakeholders collaborate because they are pursuing a mutually beneficial outcome, but while collaborating they draw on individual and collective resources in order to influence the outcome of the process to their own advantage.

Collaboration therefore provides a structure for the governance of the stakeholders involved in the production, delivery and management of the destination brand. The tourism literature shows consistency in considering the process of branding a destination as a collaborative effort (Blain, 2001; Deslandes, 2003; Im, 2003; Kaplanidou & Vogt, 2003; Morgan *et al.*, 2002, 2003; Morrison & Anderson, 2002). Collaboration in the context of destination branding is directed to the creation of a shared image of the destination through the achievement of a high degree of inclusiveness in the process (Blain, 2001). A shared image across all stakeholders is important to ensure consistency in the image delivered. An image built on shared destination attributes enhances the marketing efforts of a destination (Cai, 2002). Similarly, the sustainability of the positioning of the brand is related not only to the message delivered but also to the degree of shared meaning contained in the message (Foley & Fahy, 2004). Consistency achieved through shared meanings falls into the conceptualisation of brand equity (Aaker, 1991; Keller, 1993a), which plays a critical role in the evaluation of the effectiveness of a brand. In fact, as Keller (1998: 166) emphasised, cohesiveness is related to the success of a brand and it 'depends on the extent to which the brand elements are consistent'.

The need for stakeholders' participation in destination branding and the more general importance of participative processes in tourism planning is widely recognised and accepted (de Araujo & Bramwell, 2000; Gunn & Var, 2002; Jamal & Getz, 1995; Keogh, 1990; Laws *et al.*, 2002; Marcouiller, 1997; Reed, 1997; Sautter & Leisen, 1999; Simmons, 1994; Timothy, 1998). It has been also underlined that it is pointless to plan for tourism without at the same time taking into account the detailed needs of all stakeholders in the area (Laws, 1995). Specifically, the destination brand has been conceptualised as the positive outcome of the achievement of unity and collaboration among stakeholders (Prideaux & Cooper, 2002). In this context of collaboration, the 'collective phenomenon' of destination branding has also been described as a 'highly complex and politicised activity' (Dinnie,

2002; Morgan *et al.*, 2003: 286). The normative need for inclusiveness of multiple stakeholders with different and sometimes conflicting interests within this process should be balanced by an understanding of how stakeholders push for their own objectives.

Power in Tourism

Power has been described as a central characteristic of a collaborative process (Gray, 1989; Margerum, 2002; Shin, 2006; Soliman, 2001; Wood & Gray, 1991). The significance of power in destination branding can be analysed from two different though complementary perspectives. On one hand, adopting a conceptualisation of power as the ability to exert intentional influence (Siu, 1985) is linked to the importance that images portrayed through the destination brand have in shaping the preferences and behaviour of tourists. Under this Foucauldian (1980) point of view, the destination brand has the potential to exert power over the tourists, consistent with Castells' (1997: 359) observation that 'power lies in the codes of information and in the images of representation around which societies organise their institutions, and people build their lives and decide their behaviour. The sites of this power are people's minds'. Therefore, the recognition of the impact that brand images and brand associations have on the mind of the consumers (Keller, 1993a) makes it critical for the heterogeneous stakeholders of a tourism destination to decide, in a struggle for power, whose values – and therefore whose political agenda – is reflected by the destination brand. On the other hand, because of the critical relevance that brands have in consumers' minds, the multi-stakeholder collaborative process of destination branding is again a critical context for the understanding of power. Within this context, power can be understood in terms of the resources that a stakeholder or a group of stakeholders invests in the process of destination branding, trying to make another person or group do something they would otherwise not do (Galbraith, 1983), and sometimes even getting results contrary to their intentions (Booher & Innes, 2002).

It has been observed that 'power has such a commonsense meaning that is used too often with so little seeming need for definition' (Galbraith, 1983: 2). However, although the concept of power is of great importance in sociology (Giddens, 1993), there is much less agreement on its definition and the understanding of causes and consequences of it (Hanneman & Riddle, 2005). Trying to systematise the controversial debate among philosophers, sociologists and political scientists about the conceptualisation of power, Clegg (1989) characterised the body of knowledge about power in terms of an evident continuity in the way power is discussed.

From this perspective, causality and agency are considered not only basic assumptions that underlie how power is defined. but also emerge as the defining characteristics of the debate about power. It is necessary, nonetheless, to recognise that authors like Harré and Madden (1975) challenged a conception of power that implied the existence of causality and agency. These authors considered causality as being inherent, in the form of causal powers, to the structures of some things (Clegg, 1989). As Harré and Madden maintained:

> it is power particulars, particularly in the forms of generative mechanism, that produces new states of affairs. Sometimes they do this when stimulated by some environmental contingency, but sometimes all that is required is that the occasion for the manifestation of the powers of an individual be created by the removal of some constraints upon its activity. (Harré & Madden, 1975: 141)

The way research has approached the analysis of power in tourism can be characterised along two perspectives: a Foucauldian view on power characterised by the understanding of how images are a representation of power, and a planning and policy view that analyses the impact of power in multi-stakeholder decision-making processes in tourism. Tourism critical studies and image studies have conceptualised power and hegemony as a means to understand, as Foucault suggested:

> what rules of right are implemented by the relations of power in the production of discourses of truth? Or, alternatively, what type of power is susceptible of producing discourses of truth that in a society such as ours are endowed with such potent effects? (Foucault, 1980: 92)

From a Foucauldian perspective, discursive practices such as speech, text, talk, representations, writings, and cognition are the constituents of subjectivity. Identity, or 'subjectivity', is not a psychological essence that resides in the minds of individuals but is an effect of the regulatory operation of discourse (Dick & Cassell, 2004). In fact discursive practices are not neutral, but they are the outcome of a continuous power struggle in which power must be seen as a productive process, creating human subjects and their capacity to act (Butler, 1990). Foucault (1973, 1979) considered that power is driven by the *gaze*, whose objective is to gather and create information and to generate a discourse on a subject matter (Fox, 1988). Power and knowledge co-exist in this process. As Armstrong observed:

> power assumes a relationship based on some knowledge which creates and sustains it; conversely, power establishes a particular

regime of truth in which certain knowledges become admissible or possible. (Armstrong, 1983: 10)

The analysis of power from this perspective is particularly relevant for the analysis of the process of destination branding, considering that

> what tourists see, experience, and learn about cultures they visit is often conditioned by existing structures of image representation and interpretation of cultural others, which can re-affirm stereotypes rather than break them down. (Andsager & Drzewiecka, 2002: 402).

The construction of a gaze through imagery played a central role in how Urry (1990) conceptualised the way tourists produce and consume the space. Urry (1992: 184) proposed that 'different tourist gazes involve particular processes by which the collective memory of a society is organised and reproduced'. These gazes are

> ...self consciously organised by professionals. These include the writers of travel books and guides, travel agents, hotel owners and designers, photographers, tour operators, travel programs on TV, tourism development officers, and so on.' (Urry 1992: 173)

The socially constructed image of place is therefore built around desirable themes, selected to fit the truth of the organic images of a destination with the need to fulfil a political agenda. As Gunn (1988b) observes, the malleability of inorganic images, as opposed to the rigidness of organic images, makes them the object through which destination marketers are able to shape and create the image of the destination. Emphasising the power of images from a Foucauldian perspective, McClure observes that nature has been explicitly chosen as the lead character in the history of the development of tourism in New Zealand. Nonetheless, the transformation of New Zealand from 'a boring expanse of green hills dotted with sheep' (McClure, 2004: 285) into a successful tourism destination was possible through a politically driven marketing reconstruction of the image of the country through the brand '100% Pure New Zealand'.

Moreover, in considering power in tourism from this image-centred perspective, it is argued that 'tourism processes manifest power as they mirror and reinforce the distribution of power in society, operating as mechanisms whereby inequalities are articulated and validated' (Morgan & Pritchard, 1998: 7). Within this perspective, the 'cultural brokers of tourism' (Dann, 1996: 61) are described as playing a critical role in portraying a destination and its people through moulding, manipulating and even creating the predispositions and motives of the tourists. The conflictive interactions among hosts and guests (Brunt & Courtney, 1999; Fredline & Faulkner, 2000; Smith, 1989; Smith & Brent, 2001) have been portrayed in terms of power differentials. As Wearing and Wearing maintain, 'the power differentials between western tourists and the dominant discourses of their culture and those of the host culture can mean that the tourist may merely impose his/her ideas onto the host culture' (Wearing & Wearing, 1996: 239). The creation of tourism images as a way to express power is consistent with an argument that is central to Said's conceptualization of language and representations as demonstrations of political, intellectual, cultural and moral power. An example of this is the development of the concept of orientalism. As Said maintained:

> Orientalism is not a mere political subject matter or field that is reflected passively by culture, scholarship, or institutions; nor is it a large and diffuse collection of texts about the Orient; nor is it representative and expressive of some nefarious 'Western' imperialist plot to hold down the 'Oriental' world. It is rather a distribution of geopolitical awareness into aesthetic, scholarly, economic, sociological, historical and philological texts; it is an elaboration not only of a basic geographical distinction (the world is made up of two unequal halves, Orient and Occident) but also of a whole series of 'interests' which, by such means as scholarly discovery, philological reconstruction, psychological analysis, landscape and socio-logical description, it not only creates but also maintains; it is, rather than expresses, a certain will or intention to understand, in some cases to control, manipulate, even to incorporate, what is a manifestly different (or alternative and novel) world. (Said, 1978: 13)

The example reveals the use of particular language that can be charac-terised as language of power. This language, and the images attached to it, is specifically targeted at an audience and constructed in order to shape the perception and the images constructed by that audience. As Morgan and Pritchard observe:

> market segmentation is not simply a highly sophisticated marketing technique – seen from a different perspective, it is a process under-pinned by power relationship in just the same way as is the entire image creation process. (Morgan & Pritchard, 1998: 119)

Using hyperbole, Cheong and Miller (2000: 372) argue that 'power is every-where in tourism'. In fact, based on Lasswell's (1936) conceptualisation of politics in terms of power, Hall characterises tourism as a political arena in which decisions are political processes that involve 'the values of actors (indi-viduals, interests groups and public and private organisations) in a struggle for power' (Hall, 1994: 3). Within this scenario, multi-stakeholder decision-making processes, like the process of destination branding, are never an innocent or neutral activity (Healey, 1997, 2003). In fact, it has been argued (Treuren & Lane, 2003) that the stakeholders of a tourism destination do not face decision-making processes such as destination branding. They do not adopt a deterministic perspective that aims to optimise returns and minimise opportunity costs (Jenkins & Hall, 1997), nor can the decision-making process within a tourism destination be explained in normative terms where decisions are the outcome of a rational process of selection among different options (French *et al.*, 2000). The complexity of tourism systems (Mill & Morrison, 2002) impacts on how decision-making processes are conceptualised. In particular, the tourism system has recently been described (Farrell & Twining-Ward, 2004) as a panarchy, a hierarchical nesting of one system level within another, where founding components structure the system from the bottom up (Gunderson & Holling, 2002). Using the concept of panarchy, the tourism system can be described as structured around:

> a core [that] generally consists of an assemblage of structures, goods, services, and resources directly contributing to the sector, the comprehensive tourism system includes significant social, economic, geological, and ecological components, along with processes and functions that complement its totality and are essential to its sustainability. In the panarchy, the lower levels are semi-autonomous, facilitating some connection and transfer to the level above which is slower moving and largely unaffected by many lower level disturbances. However, small changes in one level may occasionally have unpredictable, sometimes profound effects on other parts of the same system level, triggering a cascade of repercussions which may be significantly greater than the initial disturbance.' (Farrell & Twining-Ward, 2004: 279)

These characteristics of tourism determine that in every decision-making process, such as the branding of the destination, a variety of different interest groups operating at different levels tend to influence the outcomes of the process (Hall & Jenkins, 1995). Within this context of power tensions, it is critical to assess and determine which groups are to be legitimately involved in tourism policymaking (de Araujo & Bramwell, 2000). Although it has been observed that the centrality of power for the explanation of the dynamics that govern the inter-relationships among players in the tourism industries is considered still to be 'relatively peripheral' (Morgan & Pritchard, 1999: 10), it is clear that decision-making in tourism involves a power struggle among the values of individuals, interest groups, and public and private organisations (Hall, 1994). Considering that the participants in the process of destination branding aim to convey, through the destination brand, an image of the destination that matches their own individual interests and perspectives, decision-making processes in tourism are better explained by considering:

> ongoing tension between rational, technocratic and optimising decision-making within the various levels of decision-making, and the political and symbolic context in which these decisions are made and implemented. [...] process[es] need to be understood as an ongoing reconciliation between the various and sometimes mutually exclusive interests of firm and industry profitability, community legitimacy, as well as industry and environmental sustainability. (Treuren & Lane, 2003: 2)

Social Network Power in Destination Branding

While in the strategic management literature collaboration among different firms is related to the ability of the firms to compete in an increasingly complex business environment (Bleeke & Ernst, 1993; Shuman *et al.*, 2001), collaboration in tourism, and specifically in destination branding, is a not a strategic alternative for the destination but is essential and inherent to the nature of the tourism destination. As observed by Fyall and Garrod (2005), tourism marketing has to shift its focus from the understanding of how individual organisations compete to the focus on how inter-organisational collaborative domains emerge in order to face marketing related problems. Under this new paradigm, collaboration represents a governance mechanism, required to manage the destination brand that is conceptualised as a common resource belonging and impacting on the whole tourism destination. It is also argued that the destination brand is a powerful mechanism of influence over the behaviour of tourists, and the process of destination branding is therefore an appropriate context to reveal how stakeholder power is exerted in order to influence how the inorganic images of the destination are shaped.

From a theoretical viewpoint, the social nature of destination branding as a collective phenomenon allows the understanding of how power is used in the branding process under a social network perspective. The social network perspective appears to be particularly appropriate for the study of how images are formed and how they are an expression of power over somebody or something. In fact, Foucault recommends that:

> power must be analysed as something which circulates, or rather as something which only functions in the form of a chain. It is never localised here or there, never in anybody's hands, never appropriated

as a commodity or a piece of wealth. Power is employed and exercised through a net-like organisation. (Foucault, 1980: 95)

The existence of individual interests central to the conceptualisation of power as agency (Weber, 1947), and expression of the material separation and adversarial relationship (Adams, 2002), has to be reconsidered within destination branding and transformed into what Arendt (1958: 182) calls *inter-est*, that is, 'something [...] which lies between people and therefore can relate and bind them together'. The destination brand is therefore the outcome of a process in which individual interests have been mitigated and/or exalted through the use of network power in order to produce an *inter-est* that aims to create a collaborative advantage (Huxham, 1996).

The definition of network power as the 'shared ability of linked agents to alter their environment in ways advantageous to these agents individually and collectively' (Booher & Innes, 2002: 225) must be reconsidered under the perspective of the Arendtian *inter-est*. In fact, while the patterns and the quality of the connections that characterise a social network are critical indicators of the ability of subjects to promote or protect particular interests (Dahl, 1961; Gilchrist, 2004), the need for consistency in the delivery of the destination brand requires the destination branding process to reach a high degree of inclusiveness among the tourism destination stakeholders. Despite the consideration that a process such as destination branding can be considered collaborative, even if takes place only among 'the most powerful or influential stakeholders [...] whatever their power' (Wood & Gray, 1991: 155), the marketing implications of inconsistencies (de Chernatony & Harris, 2000; Speak, 1996) in destination brand delivery for the destination as a whole require a reformulation of the definition of network power in destination branding. Booher and Innes' (2002) definition, which has particular relevance in urban planning studies, can be transposed into destination branding by understanding the implications carried by concept of brand equity (Aaker, 1991; Keller, 1993a) and the closely related concept of destination brand equity (Konecnik, 2006). Network power can be conceptualised as the shared ability of linked stakeholders of a tourism destination to influence the process of destination branding towards an outcome (the destination brand) that reflects the values and business agenda of the largest possible number of tourism destination stakeholders.

The use of a social network perspective on power in destination branding carries different implications for the management of a tourism destination. In fact, the asymmetry that characterises how individual destination stakeholders can exert power in a destination branding process represents a threat to the destination brand equity (Konecnik & Gartner, 2007). While individual operators might have enough resources to shape the destination brand to fit their objectives and interests, the symbiotic dependency that exists between individual stakeholders within a tourism destination might affect the overall performance of the tourism destination. Considering this point, Sautter and Leisen (1999: 313) argue that the role of destination management organisations is to 'maximise positive returns to a community's overall growth', reinforcing Pike's consideration that the role of a destination management organisation is to contribute to the governance of the destination by 'encapsulating in a destination brand the essence or spirit of a multi-attributed destination representative of a group of sellers as well as a host community' (Pike, 2004: 92). Destination management organisations should conceptualise their role as facilitators in the process of linking individual stakeholders together with the objective of creating links and bounds amongst the destination stakeholders. The incorporation of a social network approach to leverage stakeholder power in destination branding towards the creation of positive destination brand equity requires destination management organisations to systematically use social network analysis (Galaskiewicz, 1996; Galaskiewicz & Wasserman, 1993) as a diagnostic tool. Mapping the destination stakeholders in this way allows an understanding of how the centrality of one or more stakeholders within the destination enhances or reduces the ability of the destination management organisation to work towards the creation of the Arendtian *inter-est*.

Conclusion

Beginning from a conceptualisation of destination branding as a collaborative effort amongst stakeholders, this chapter proposed the incorporation of the concept of network power within the discussion on destination branding. The image portrayed by the destination brand has an impact both on the destination as a whole and on individual stakeholders within the destination. It is therefore to be expected that single stakeholders will exert influence on the destination branding process in order to shape its outcome in favour of their individual objectives. However, the need to preserve and enhance destination brand equity as the common good of the destination imposes the need to transform the individual interests into Arendt's *inter-est*. With this objective, network power has here been conceptualised as the shared ability of linked stakeholders of a tourism destination to influence the process of destination branding towards an outcome (the destination brand) that reflects the values and business agendas of the largest possible number of tourism destination stakeholders.

Part 2
Quantitative Approaches to Tourism Network Analysis

Chapter 11

Issues in Quantitative Network Analysis

Introduction

In this chapter we begin with an introduction to the measurement issues involved in the development and analysis of a quantitative network analysis project. We then continue by outlining the mathematical basis for network analysis. One of the reasons for the use of network theory in tourism research is that network methodology interacts strongly with and indeed is embedded in theory and guides data collection towards collection of data about relationships (see Chapter 3).

This chapter can provide only an overview and introduction to a complex and deep literature. The reader is also referred to a number of prior papers that have examined and summarised the literature on network data collection and analysis (Breiger, 2004; Haythornthwaite, 1996; Marsden, 1990). Others provide more topic-specific examinations in subject areas such as health which, however, may be relevant for design of tourism research (Brinton *et al.*, 1998; O'Reilly, 1988). Useful handbooks on network analysis include those by Scott (2000) and Wassermann and Faust (1994).

A final point about network research is that its widespread use requires high speed computers and software packages specifically designed for network analysis. A number of such packages are available including a number that are free or inexpensive. It is recommended that programs such as UCINET (Borgatti *et al.*, 1999) are investigated and used for network research. These programs are relatively easy to understand and master.

A Review of Tourism Network Methodology

There are two theoretical approaches to the concept of a network which lead to a requirement for different types of data to be collected. A network may be a sensitising metaphor or a conceptual representation of social structure. In both cases the behaviour of actors is seen as a function of their varying positions within a social network (Mizruchi, 1994). However, these two approaches lead to different research traditions (Wellman,

1988). The first is qualitative in nature, using techniques derived from anthropology and ethnography (Pavlovich, 2001; Tinsley & Lynch, 2001) while the other is quantitative and uses social network analysis techniques based on mathematical algorithms (Pforr, 2002). In the wider literature there have been extensive methodological developments in the domain of quantitative analysis (Wellman, 1988) and this is the focus of this chapter. While in both approaches the objects of study are the transactional content, the nature of the links and the resultant structural characteristics of the network, the type of data collected is quite different, with metrics used in one case and 'thick description' used in the other.

What data do you need to collect?

Social network analysis studies may be characterised in terms of the types of social objects being studied, the transactional content, the nature of the links and the resultant structural characteristics examined (Knoke & Kuklinski, 1991). Before collecting data, a network researcher must decide on the most relevant type of social organisation to be studied, and the units within that social organisation that will comprise the network nodes. Possible social organisations for study include individuals, aggregates of individuals, organisations, classes and strata, communities, and nations (Cross *et al.*, 2002).

The ideal study examines a whole network and describes the ties that all members of a domain maintain. This approach requires responses from all members of a particular domain and therefore may be time-consuming or expensive. The number of possible ties is equal to the size of the population, n, times n -1. For a population size of 20, there are 380 potential links for each relationship. Thus, researchers are often limited by the number of actors to include or by the number of relationships they can reasonably study.

A second type of study examines the relationship that a particular actor maintains with others, called an egocentric network. Egocentric networks build a picture of a typical actor in any particular domain and show how many and what type of relationships such actors have to others. This approach is particularly useful when the population is large, or the boundaries of the population are hard to define. This approach was used by Granovetter (1973) to examine how people find jobs. Other studies have adopted a focus intermediate between the individual and the population. Most often these are dyads, but triads and even larger subsets are also studied.

It is important to ensure that a research project is delimited by specification of a network's boundaries (Thatcher, 1998). This in some ways parallels the general problem of defining the population to which research results are to be generalised. It is of special importance in network studies, however, since analyses focus explicitly on interdependencies among the particular units studied. Omission of pertinent elements or arbitrary delineation of boundaries can lead to misleading results. Determining a boundary for a destination network study may be done by alternatively focusing on the organisations, their relations or critical policy events (Pforr, 2002). The focus is on actors sharing a common goal or using actors located within geographical limits (Laumann *et al.*, 1978: 460). The idea of focusing on actors within a geographical area naturally overlaps with the study of destinations, clusters or industrial districts as these also have a geographical basis.

The identification of actors within a destination network within constraints of time and money requires some criteria for selection. One common method is to distinguish between actors on the basis of their degree of influence, and leads to the use of key stakeholders. Snowball sampling may be used to follow relationships and identify other nodes (Atkinson & Flint, 2001). Various methods and approaches have been used to identify key stakeholders, including the position approach, reputation method, decision method or participation/relational methods (Knoke & Kuklinski, 1991; Thatcher, 1998; Tichy *et al.*, 1979). Each of these approaches has strengths and weaknesses, as discussed in Tichy *et al.* (1979) (see Table 11.1).

Table 11.1 Methods used to identify key stakeholders

Method	*Actors identified*
Positional methods	Persons or organizations occupying the key roles in the analytic system, such as the elected or executive positions in major economic and political units
Decisional methods	Actors that participate in making or influencing the collectively binding decisions for the system as a whole
Reputation methods	Actors widely believed by knowledgeable observers to have actual or potential power to 'move and shake' the system.
Participation/ Relational methods	Actors who maintain important political relationships with other system members, who were not uncovered during queries about elites power reputations.

Source: Knoke & Kuklinski (1991)

Relationship content

The transactional content of a relationship is what is exchanged when two actors are linked. For instance, two employees may exchange information or affect. Different types of transactional contents can be distinguished, such as exchange of affect (liking, friendship), exchange of influence or power, exchange of information, and exchange of resources, goods or services; conflict links are occasionally included as well. Szarka (1990) for example discusses three types of networks among small business – social, communication and exchange. Lynch (2000) has studied social and business relationships in the homestay sector. Efforts at empirical typology for types of interpersonal relations (Burt, 1983) suggest that they vary along dimensions of tie strength, frequency of contact, and role relationships (a contrast of kinship versus workplace contact).

The particular content to be examined depends on what is being studied, and careful choices must be made in deciding on the relationships to ask about in order to study the particular issue. When it is not known which relationships are most relevant, or when all interactions between pairs are to be examined, it may be necessary to ask about a large number of relationships.

Direction of the links

The nature of the linkages between pairs of individuals can be described in several ways, such as intensity or reciprocity. *Intensity* is the strength of the relation as indicated by the degree to which individuals honour obligations or forego personal costs to carry out obligations (Mitchell, 1969), or by the number of contacts in a unit of time. *Reciprocity* is the degree to which individuals report the same (or similar) intensities with each other for a content area. When information is passed from one person to another, it flows in a certain direction. Phrasing questions about relationships to tease out who gives what to whom can indicate the direction in which information or resources flow. Relationships can also be undirected. An undirected relationship is one in which the direction of flow is either not measured or is considered not relevant; for example, joint membership on a board, or co-authorship on a paper. An undirected relationship can be measured in both directions; for example, actor A 'talks to' actor B. Another attribute of relationships is strength. *Strength* refers to the intensity of a relationship; for example, a relationship in which a large number of resources are exchanged or in which actors meet and exchange information frequently is stronger than a relationship in which few goods are exchanged or in which information is exchanged infrequently.

An example

In a recent study of tourism organisations in Queensland Australia, the initial sampling of organisations was based on identification of key tourism sector stakeholders using the reputation method. Based on initial discussions with staff from Tourism Queensland (the State Government Tourism Office) and further snowball sampling, 22 key organisations were identified and interviewed. This method allows prioritisation of key stakeholders for contact based on the number of times each was mentioned by other organisations. It also allowed the number of interviews in each region to be reduced to a manageable number. Efforts were made to obtain comments from each of these organisations about each of the other 21 organisations as well as other organisations considered to be important. As a result, during the study several organisations from outside the region were identified and included but not interviewed due to time constraints. While the study may be seen to use small numbers of respondents, these respondents were perceived as the key stakeholders in the region.

Each organisation was interviewed face to face with the interviews taking around one hour. A written questionnaire was used and in particular respondents were asked a series of semi-structured questions concerning the organisations they had relationships with, the frequency of interaction and the type of interaction. Data was collected using predetermined code frames. A number of open-ended questions were also asked during the interview.

Structural properties

There are a number of structural properties of networks that have been described in the literature and these are commonly provided as output from network software programs. Network size is a basic indicator of interest and measures the number of direct ties involving individual units, which may measure the degree of integration in a network. Haythornthwaite (1996) discusses five principles that network analysts use to examine an actor's network:

- *Cohesion;* grouping actors according to strong common relationships with each other.
- *Structural equivalence;* grouping actors according to similarity in relations with others.
- *Prominence;* indicating who is 'in charge'.

- *Range;* indicating the extent of an actor's network.
- *Brokerage;* indicating bridging connections to other networks.

Cohesiveness describes attributes of the whole network, indicating the presence of strong relationships among network members, and also the likelihood of their having access to the same information or resources. Overall measures of cohesion, such as *density* and *centralisation,* indicate the extent to which all members of a population interact with all other members. In addition, by identifying areas of a graph that show a higher degree of connectedness, network structures such as *clusters* and *cliques* can be revealed. *Clusters* are subgroups of highly interconnected actors. When fully interconnected, these clusters are known as *cliques.* Within cliques, members can reach each other directly in one step without going through an intermediary.

Density measures the relative number of ties in the network that link actors together and is calculated as a ratio of the number of relationships that exist in the network (stakeholder environment), compared with the total number of possible ties if each network member were tied to every other member. A complete network is one in which all possible ties exist (Rowley, 1997). *Centrality* is a related measure and can inform how decisions are made and how information flows around a network. A node with high *betweenness* has great influence over what flows in the network. Betweenness is the position of an actor between cliques or groups. It can be conceptualised as the extent to which a stakeholder has potential control over other stakeholders' access to other parts of the network.

Similarly *closeness* measures the path length between actors. Some actors are close to everyone else. They are in an excellent position to monitor the information flow in the network – they have the best visibility into what is happening in the network. Boundary spanners are more central than their immediate neighbours whose connections are only local, within their immediate cluster. Boundary spanners are well positioned to be innovators, since they have access to ideas and information flowing in other clusters. They are in a position to combine different ideas and knowledge into new products and services.

Using these principles, network analysts can explore networks to determine what subgroups of interconnected actors exist (Alba, 1982). Identification of cliques and clusters of nodes within the network may reveal the operation of 'a virtual company'. *Structural equivalence* identifies actors with similar roles. Actors are considered to be structurally equivalent if they fill the same role with respect to members of the same network.

Collecting the Data: Traditional and New Techniques (Data Sources)

Once the parameters of the research have been identified, data collection can proceed. The data sources that may be used for social network analysis are as diverse as in other areas of social research, although surveys/questionnaires or self-reports are the predominant research method used. Most often such data are obtained with single-item questions that ask a respondent to enumerate those individuals with whom they have direct ties of a specified kind. One recent tourism study used archival analysis to identify respondents, followed by a mail questionnaire (Pforr, 2002). Other methods include diaries, electronic traces, observation, informants, and experiments. Problems with data collection are similar to those in other areas of the social sciences, although particular problems may be experienced in quantifying relationships and also because different types of relationships may be intertwined (Thatcher, 1998).

When surveys and questionnaires are used to study inter-organisational relationships, problems of respondent selection arise due to specialisation within organisations. Most studies select only one agent to report on an organisation's ties to all other organisations, but it is plausible to expect that the quality of such reports might be better for those kinds of relations that involve the informant's own activities. The quality of network data obtained by surveys and questionnaires is far from perfect, and gathering such data often requires substantial research budgets. Archival sources of various kinds are inexpensive, and advantageous for studying social networks in the past or in which units are otherwise inaccessible.

Other methods of assembling network data have been used less often. The social anthropologists who were early contributors to development of the network orientation tended to rely on observational methods of data collection, which may yield greater descriptive accuracy but are very time-consuming. It is possible to recruit participants to keep diaries of their contacts over a period of time, or even to gather data from internet conversations or other electronic communications.

The Mathematical Basis for Network Analysis

Elementary graph theory

A very familiar representation of a network is as a drawing in which a number of points are connected by some lines. Points and lines may have numerous synonyms, some of them used more frequently in different disciplines, so a point is also a vertex, a node and an actor, while a line is also called an edge, arc, link, relationship or tie. In the mathematical abstraction of a diagram consisting of points and lines connecting them, the object known as graph G, is, formally, an ordered pair of disjointed sets *(V,E)* where $V = \{v_1,...v_n\}$ is the set of vertices and $E = \{(v_1,u_1),..., (v_i,u_j)\}$ is the set of arcs. If the elements of E are ordered pairs (i.e.: (v_i, v_j) ? (v_j, v_i), the line has a direction), the graph is said to be directed; it is undirected otherwise. (Diestel, 2005; Bollobás, 1998). Each line can be assigned a numeric value called *weight*; it can represent, for example, a distance, a time or a cost (the graph is then said weighted). There is no unique way of drawing a graph; the geometric positions of points and lines have no special meaning. As noted, the properties of a graph depend on its form and on the relationships among its elements. The following is a list of the main definitions and of the principal parameters characterising the topology of a graph (the appendix at the end of this chapter contains a list of the mathematical expressions for the main network metrics):

- *Order* is the number of vertices of a graph, *size* is the number of edges (often, however, the term size is used to indicate the number of vertices in a network).
- A graph is *complete* if each vertex is connected to every other.
- A *subgraph* of a graph G is a graph whose vertex and edge sets are subsets of those of G.
- A *clique* is a complete subgraph of a graph.
- A *bipartite* graph is a graph which can be divided into two disjointed parts A and B so that any edge connects a vertex in A with a vertex in B, but there are no edges connecting vertices within the same part.
- The *density* of a graph is the ratio between its size and the maximum possible number of edges that a graph may have (the size of a complete graph of the same order).
- The *degree* of a vertex is the number of edges that connects it to other vertices (if the graph is directed, the degree can be distinguished in *indegree* and *outdegree*, meaning the number of arcs arriving to or departing from the vertex).
- In a *regular* graph all vertices have the same degree.
- The *degree* distribution is a function which gives the probability (the number) that any vertex has a certain degree.
- A *path* is a series of consecutive edges connecting the initial and the final vertices of the path; if any two of vertices in a graph have a path between them, the graph is *connected*.
- The *distance* between two vertices is the shortest path of edges connecting them (in case of a weighted graph, the distance is the path whose total weight is the lowest possible).
- The *diameter* of a graph is the longest distance (the maximum shortest path) existing between any two vertices in the graph.

- *Neighbours* (or *first neighbours*) of a vertex are all the other vertices directly connected with it; the *neighbourhood* of a vertex is the subgraph containing it and all its first neighbours.
- The *clustering coefficient* of a vertex is a measure of local density. It is calculated as the ratio between the actual number of edges connecting the neighbourhood of a vertex and the maximum possible number of edges of that neighbourhood.
- The *closeness* of a vertex is the reciprocal of the sum of geodesic distances to all other vertices in the graph; it can be interpreted as a measure of how long it will take information to spread from a given vertex to others in the network.
- The *betweenness* is a centrality measure of a vertex within a graph. It is normally calculated as the fraction of shortest paths between node pairs that pass through the one of interest. In social sciences, the betweenness is a measure of the importance or of the influence of an actor in a group.

Maximums, minimums or averages of these quantities, calculated over all the vertices or the edges of the graph, characterise its topology. The adjective *global* is also used to indicate an overall average.

A finite graph (with a finite number of vertices) can be represented by its *adjacency matrix*: an $n \times n$ matrix whose entry in row i and column j gives the number of edges from the *i-th* to the *j-th* vertex. For an undirected unweighted network the adjacency matrix is symmetrical and boolean (it contains only 0-1 values). The relation between a graph and its adjacency matrix is *biunivocal*; this means that a graph is fully identified by its adjacency matrix and vice versa. The methods and the techniques of linear algebra (Lang, 1970) can thus be employed for the study and the calculation of graph properties.

An example of practical usage of the adjacency matrix A_G of a graph G is in the calculation of the paths of a graph. The number of paths of length n from v_i to v_j is the (i, j) entry of the matrix obtained by raising A_G to the *n-th* power: . If A_G is raised to a power m so that all the elements a_{ij} of A_G are positive, m is the diameter of G.

A square matrix of order n has n eigenvalues (Lang, 1970); this set is called spectrum of the matrix (and of the graph). Spectral theory studies the relationships between the properties of the graph and its adjacency matrix and can be used to analyse the topological properties of a graph (Seary & Richards, 2003). Eigenvalues are, in fact, invariant to the order of nodes in the graph. In particular, the largest eigenvalue is connected with the density of links, and the second smallest is a measure of the

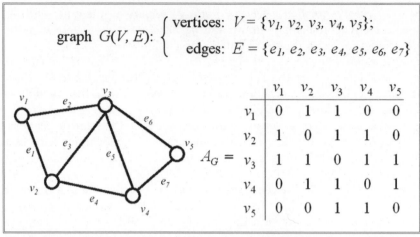

Figure 11.1 A graph $G(V, E)$ with the corresponding adjacency matrix A_G.

compactness (algebraic connectivity) of a graph. A large second eigenvalue indicates a compact graph, whereas a small eigenvalue implies an elongated topology.

Positive eigenvalues are associated with clustering of interconnected nodes and negative ones with partitioning of the network into sets of nodes that have similar patterns of connections with nodes in other sets, but few connections amongst themselves.

The eigenvectors of a graph have been also been used to give a measure of centrality of a vertex (Bonacich, 1972). The *i-th* component of the eigenvector corresponding to the largest eigenvalue gives the centrality score of the *i-th* node in the network. The idea is that connections to nodes having a high score contribute more to the score of the node itself. It is a variation of the betweenness measure. Google's PageRank (Brin & Page, 1998) is a variation of the eigenvector centrality measure.

Network models

The first mathematical model, used for many years to describe many kinds of networks, is that of Erdõs and Rényi (1959, 1960, 1961). In the ER model, a graph is a set of n nodes connected two at a time with probability p: $G(n, p)$. The random distribution of the nodes degrees k (the number of connections) follows, for large n, a Poisson law with a peak $<k>$:

$$P(k) \approx \langle k \rangle^k e^{-\langle k \rangle} / k!$$

This implies that most vertices have about the same number of links (the average degree $\langle k \rangle$), while nodes with a degree that deviate significantly from the average are extremely rare. The *tail* (high k region) of the degree distribution $P(k)$ decreases exponentially.

The clustering coefficient C depends on the average degree of the network $\langle k \rangle$; it is therefore constant for a given graph. The mean path length L is proportional to the logarithm of the network order, $L \propto \log n$. An ER graph is in equilibrium, the number of vertices is fixed. A possible evolution is obtained by varying the connection probability. One of the most interesting results of this model is that a number of peculiar characteristics depend strongly on special values of p.

Let us consider, for example, an ER graph $G(n, p)$ and let the probability $p = c/n$ for some positive number c. For values $c<1$ it is possible to identify in G several small connected components whose order is proportional to $\log n$; at $c = 1$ a sudden and dramatic change takes place in G: a giant connected component of order $S = n^{2/3}$ appears. The order of this giant component is a constant fraction of all the vertices of the graph; the other components have much smaller size ($S \propto \log n$).

At these boundaries, a phase transition occurs similar, for example, to the one found in studying percolation phenomena (Callaway *et al.*, 2000).

The vertex degree distribution $P(k)$, the clustering coefficient C and the average length L of the shortest paths between two vertices have been found to be characterising parameters for the topology of a network (Boccaletti *et al.*, 2006; Watts, 2004; Newman, 2003; Albert & Barabási, 2002; Dorogovtsev & Mendes, 2002).

At the end of the 1990s, empirical studies (see Chapter 2) confirmed that, in many cases, ER random graphs are quite different from *real* world networks. Following these, numerous investigations have been done and a great variety of theoretical works have been published.

The first evolution of the ER model is the one proposed by Watts and Strogatz (1998).

They noticed that a number of examples (the gene regulatory network, the network of collaboration among film actors and the electrical power distribution network) exhibited clustering coefficients significantly higher and mean shortest path lengths lower than expected.

Small-world (SW) networks, as they call them borrowing from Milgram (1967), arise as the result of a random replacing (rewiring) of a fraction p of the links of a regular lattice with new random connections. In this evolutionary process they position themselves between the two limiting cases of a regular lattice ($p = 0$) and a random graph ($p = 1$).

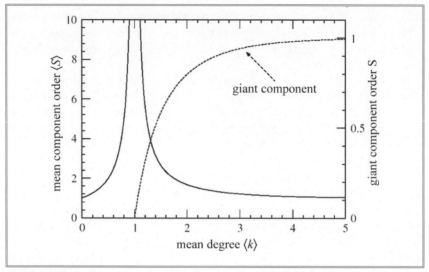

Figure 11.2 Average size (order) of the components in an ER graph and of the giant component (after Newman, 2003)

A SW network is still characterised by a poissonian degree distribution, the local neighbourhood is preserved and the diameter increases logarithmically with the number of vertices n (Amaral *et al.*, 2000; Watts & Strogatz, 1998). This is why they are called small-world networks: it is possible, on the average, to connect any two vertices through just a few links.

The analysis of other real world networks (Internet routers and web pages in particular) led the Faloutsos brothers (1999) and Barabási and Albert (1999) to the discovery that such systems had a peculiar characteristic. Their degree distribution approximates a power law: $P(k) \propto k^{-\gamma}$. The distribution is largely uneven; there is no characteristic mean nodal degree (the mean of a poissonian ER or SW distribution). Some (few) nodes act as very connected hubs, having a very large number of ties, while the majority of nodes have a small number of links. The absence of a characteristic average degree $\langle k \rangle$, the characteristic *scale* of the network, has gained these networks the name *scale-free* (SF).

SF networks are dynamic systems; they grow with the addition of new nodes and new links that are not distributed randomly, but follow specific mechanisms. The most commonly invoked is a *preferential attachment*, in which a new node has a higher probability to attach to one of the most connected ones. Deviations from pure power -laws (kinks, cut-offs etc.), which can be observed, are generally explained by introducing corrections

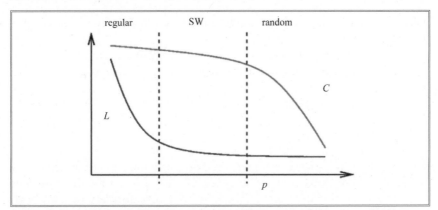

Figure 11.3 Average path lengths and clustering coefficients as function of the rewiring probability p

or nonlinear terms in the expression of the preferential attachment, by considering the limitations given by the finiteness of the network size or

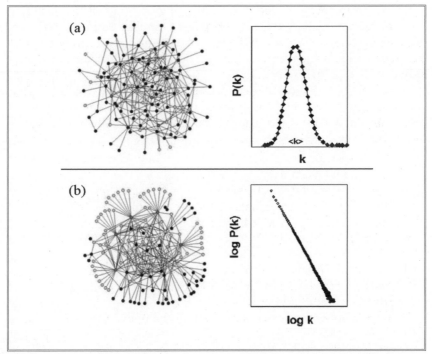

Figure 11.4 A random network (a) and a scale-free network (b) with their respective degree distributions

by assigning special properties (age, fitness or attractiveness) to some of the network actors.

The network thus created does not have an intrinsic modularity (the clustering coefficient C is independent of the degree k). Scale-free networks with degree exponents $2 < \gamma < 3$, values found in most real cases, have very small average path lengths (typically $L \propto \log \log n$, significantly shorter than the one typical of SW networks).

Many complex networks, mainly if they represent social or biological systems, exhibit multiple characteristics such as modularity, high local densities and scale-free topology. The nodes are part of highly clustered areas, with few hubs that are responsible for connecting the different neighbourhoods. Hierarchical models of network formation have been devised (Ravasz & Barabási, 2003) assuming that clusters mix in an iterative way and create a hierarchical structure. The resulting network has a power-law degree distribution and a large average clustering coefficient. More importantly, C scales following a power-law $C(k) \propto k^{-1}$. This latter characteristic is assumed to be the signature of a hierarchical network.

Both SF and SW networks have been revealed as very common structures among real world networks: film actors, company directors, scientific co-authorships, telephone calls, email messages, student relationships, sexual contacts, web pages, internet routers, word co-occurrences, electric power grids, train routes, electronic circuits, metabolic networks, protein interactions, ecosystem food webs and neural networks.

The topology of SW or SF networks is directly related to the peculiar characteristics of their behaviour in many occasions. The most important results obtained in this field regard processes such as (Newman, 2003):

- *Rich-get-richer effects*: the accumulation of some quantity (links, but also money or knowledge) directly coming from the preferential attachment mechanisms.
- *Robustness*: stability of the system to random removal (or failure) of randomly chosen elements, therefore a higher capacity to resist cascading failures or avalanches of breakdowns.
- *Fragility*: high sensitivity to targeted attacks to the most connected hubs.
- *Congestion factors*: in a random graph substrates tend to unity with increasing network size, indicating that these networks will become extremely congested at this limit. For SF networks, however, the congestion factor does not depend on the network size; arbitrarily

large networks can be considered without increasing their congestion level.

- *Low internal friction*: extent and speed of *disease* (viruses, but also messages, fads, beliefs, knowledge etc.) transmission are greatly improved in a SF or SW network with respect to a random ER graph; in some cases it is shown that there are no critical thresholds at all for these phenomena.

Future Directions for Network Methodological Development

In tourism network studies there has been a focus on qualitative data collection and analysis. This chapter seeks to broaden the range of tools available to study a critical part of the tourism domain. Use of quantitative techniques can provide insight to supplement 'thick description', often with little additional effort by the researcher. One criticism of network studies by qualitative researchers is that they are static and do not capture the dynamic nature of networks. However, use of quantitative techniques can provide useful evidence of changes in networks if successive waves of data collection are conducted.

Tourism networks are a new area for the application of quantitative network methods and would benefit from development of standard procedures and methods for data collection and analysis. The use of quantitative techniques is recommended to other tourism researchers for their investigations.

Appendix

Formulas useful in network analysis

A network is represented mathematically by a graph and by its associated adjacency matrix.

Definition 1: a graph G, is an ordered pair of disjointed sets (V, E) where $V = \{v_1,...v_n\}$ is the set of vertices and $E = \{(v_1,u_1),..., (v_i,u_j)\}$ is the set of arcs. If the elements of E are ordered pairs (i.e.: (v_i, v_j) ? (v_j, v_i), the line has a direction), the graph is said directed, it is undirected otherwise.

Definition 2: An adjacency matrix is an $n \times n$ matrix whose entry in row i and column j gives the number of edges from the i-th to the j-th vertex. For an undirected unweighted network the adjacency matrix is symmetrical and boolean (it contains only 0-1 values). The relation between a graph and its adjacency matrix is biunivocal; this means that a graph is fully identified by its adjacency matrix and vice versa.

Table 11.2 (see pp. 161–2) gives the mathematical expressions used to calculate the main parameters of a network.

The elements of the adjacency matrix A (*i rows* × *j columns*) are denoted: a_{ij}. The matrix is a square symmetric matrix, i.e. $i = j$ and there are no self-connections, i.e. no edges of the form (l, l); for each pair of edges (l_1, m_1) and (l_2, m_2) : $l_1 \neq l_2$ and $m_1 \neq m_2$. That is to say that the diagonal elements of A are null. We consider only unweighted graphs (all elements have value 0 or 1). For extension of the formulas to weighted networks the reader may see Barrat *et al.*, 2004; Barthélemy *et al.*, 2005; Newman, 2004; Saramäki *et al.*, 2006. More formulas and relationships can be found in Bollobás, 1998; da F. Costa *et al.*, 2005; de Nooy *et al.*, 2005; Dorogovtsev & Mendes, 2003; Godsyl & Royle, 2001; Scott, 2000; Wasserman & Faust, 1994.

Table 11.2 Mathematical expressions used to calculate the main parameters of a network

Order number of nodes, sometimes also called size	N
Size number of links	$$m = \sum_i \sum_j a_{ij}$$
Link density	$$\delta = \frac{2m}{n(n-1)}$$
Degree of a node	$$k_i = \sum_i a_{ij}$$
In-degree	$$k_i^{in} = \sum_j a_{ji}$$
Out-degree	$$k_i^{out} = \sum_j a_{ij}$$
Average path length d_{ij} is the length of the shortest path connecting nodes i and j	$$l = \frac{1}{n(n-1)} \sum_{i \neq j} d_{ij}$$
Diameter	$D = max(d_{ij})$
Clustering coefficient of node i t_i is the number of links between neighbours of i	$$C_i = \frac{2t_i}{k_i(k_i-1)}$$
Clustering coefficient of the network	$$C = \frac{1}{n} \sum_i C_i$$

Table continued on next page

Table 11.2 *continued*

Betweenness of node i D_{jl} is the number of shortest paths between j and l belonging to the neighbourhood of i, $D_{jl}(i)$ is the number of these which pass through i	$$B_i = \sum_{i \neq j \neq l} \frac{D_{jl}(i)}{D_{jl}}$$
Closeness of node i reciprocal of the sum of geodesic distances d_{ij} from node i to all other nodes (j) in the graph	$$CL_i = \frac{1}{\sum_{j \neq i} d_{ij}}$$
Local Efficiency of node i d'_{lm} is the shortest distance between any two neighbours of i	$$E_{loc,i} = \frac{1}{k_i(k_i - 1)} \sum_{l \neq m} \frac{1}{d'_{lm}}$$
Local efficiency of the network	$$E_{loc} = \frac{1}{n} \sum_i E_{loc,i}$$
Global efficiency	$$E_{glob} = \frac{1}{n(n-1)} \sum_{i \neq j} \frac{1}{d_{ij}}$$
Assortative mixing coefficient As Pearson correlation coefficient; dg_i is the degree of node i, dn_i the mean degree of its first neighbours; the standard error can be calculated by using the bootstrap method on the correlation	$$r = \frac{\sum_i (dg_i - \overline{dg})(dn_i - \overline{dn})}{\sqrt{\sum_i (dg_i - \overline{dg})^2 \sum_i (dn_i - \overline{dn})^2}}$$

Chapter 12

Visualising Tourism Networks: Connecting the Dots

Introduction

Over the past 50 years there have been a number of methods developed to visualise network information in a manner that demonstrates the important features of the network structure. Visualisation benefits a researcher as it helps communication about the data to third parties as well as facilitating exploration of the data (Brandes *et al.*, 1999). Figure 12.1 shows the two main ways that researchers represent information about patterns of ties among social actors: matrices and graphs. The result of a data collection exercise using network theory to examine 14 people is a 14 by 14 adjacency matrix of relationships, as shown in Figure 12.1(a). While such a matrix form contains useful information, it is not as useful a way of communicating information as Figure 12.1(b). Instead, Figure 12..1(a) addresses questions such as: Are all the nodes connected? Are there many or few ties among the actors? Are there

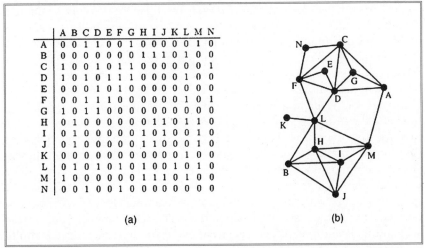

	A	B	C	D	E	F	G	H	I	J	K	L	M	N
A	0	0	1	1	0	0	1	0	0	0	0	0	1	0
B	0	0	0	0	0	0	0	1	1	1	0	1	0	0
C	1	0	0	1	0	1	1	0	0	0	0	0	0	1
D	1	0	1	0	1	1	1	0	0	0	0	1	0	0
E	0	0	0	1	0	1	0	0	0	0	0	0	0	0
F	0	0	1	1	1	0	0	0	0	0	0	1	0	1
G	1	0	1	1	0	0	0	0	0	0	0	0	0	0
H	0	1	0	0	0	0	0	0	1	1	0	1	1	0
I	0	1	0	0	0	0	0	1	0	1	0	0	1	0
J	0	1	0	0	0	0	0	1	1	0	0	0	1	0
K	0	0	0	0	0	0	0	0	0	0	0	1	0	0
L	0	1	0	1	0	1	0	1	0	0	1	0	1	0
M	1	0	0	0	0	0	0	1	1	1	0	1	0	0
N	0	0	1	0	0	1	0	0	0	0	0	0	0	0

(a)   (b)

Figure 12.1 The adjacency matrix of a political network (Reproduced from Doreian & Albert, 1989)

sub-groups or local 'clusters' of actors that are tied to one another, but not to other groups? Are there some actors with many ties, and some with few?

A good drawing can also help us to understand how a particular 'ego' (node) is 'embedded' (connected to) its 'neighbourhood' (the actors that are connected to ego, and their connections to one another) and to the larger graph (is 'ego' an 'isolate' or a 'pendant'?). By looking at 'ego' and the 'ego network' (i.e. 'neighbourhood'), we can get a sense of the structural constraints and opportunities that an actor faces and we may be better able to understand the role that an actor plays in a social structure.

There are lots of different kinds of 'graphs.' Bar-charts, pie-charts, line and trend charts, and many other things are called graphs and/or graphics. Network analysis uses (primarily) one kind of graphic display that consists of points (or nodes) to represent actors, and lines (or edges) to represent ties or relations. When sociologists borrowed this way of graphing things from the mathematicians, they renamed their graphics 'sociograms'. Mathematicians know these kinds of graphic displays by the names 'directed graphs', 'signed graphs' or simply 'graphs'. There are a number of variations on the theme of sociograms, but they all share the common feature of using a labelled circle for each actor in the population we are describing, and line segments between pairs of actors to represent the observation that a tie exists between the two.

Types of Visualisation Techniques

Network analysis techniques are used to establish the network position of stakeholders and the relationships between them using indicators such as intensity of communication, reputation or resources (Thatcher, 1998: 399). Moreno (1934) indicated that social configurations have definite structures which can be described as 'sociograms' to visualise the flow of information between organisations or individuals. This led to the development of graph theory where the relationships between nodes are represented as points and lines and the resulting patterns are described. Later developments led to the identification of groups of individuals with similar patterns of relationships (blockmodels) and to the use of statistical methods such as multidimensional scaling to transform relationships into social distance and map them in social space. A number of different techniques may be used to display the graphical data, from use of hand-drawn relational 'maps' to diagrams derived by using sophisticated statistical techniques. Apart from sociograms or graphs, other visualisation techniques include the multidimensional scaling (MDS) scattergram (Schneider, 1992), the dendrogram (Scarini, 1996) and venn diagrams (Brandes *et al.*, 1999).

This visualisation approach is particularly attractive as it compactly displays the relevant actors in a network, and how they relate to each other (Brandes *et al.*, 1999). The elements of a social network that require visualisation are the nodes, relationships, the positions of nodes and relationships in relation to one another, and sub-networks or clusters. Nodes may be represented using different shapes and colours as a means of conveying information about their characteristics. Programs such as NetDraw allow node properties to be changed easily. Relationships between nodes are usually represented by line segments in a simple graph or arrows in a directed graph. A single-headed arrow indicates the direction of a relationship, while a double-headed arrow shows mutual interaction between nodes. Also, relationships may be positive or negative, indicated with a plus or minus sign. These relationships may also have 'attributes', and colour and size may be used to indicate differences of kind and amount.

The relative position of nodes and relationships are usually drawn in a two-dimensional 'X-Y axis' space, although Mage and some other packages allow 3-dimensional rendering and rotation. In some analyses the position of a node or a relation in the space is essentially arbitrary, and it is considered that the full information about the network is contained in its list of nodes and relations. One example is the circle diagram, where the actor nodes are placed on an imaginary circle. This approach makes the pattern of lines more visible (Brandes *et al.*, 1999).

In other studies, the positions are determined by heuristics that seek to better present data. One conceptually simple heuristic for displaying relationships is the Spring Embedder technique (Eades, 1984). This is a heuristic for laying out arbitrary kinds of networks. The basic idea is to consider the nodes of the network to be repelling rings. Those nodes linked by relations are considered joined by a spring, and a positioning with low forces exerted on the rings is sought. The resultant diagram is then interpreted visually, as distances and directions provide useful information. A number of computer software packages are available to map relational data (Scott, 1996). While there is no correct approach to visualising network data, experimenting with different approaches can be useful.

Visualisation of large networks is often difficult as they contain lots of information. Often, it may be useful to simplify the data through elimination of secondary relationships, examining only one type of node or focusing in on a particular sub-network. Highlighting sub-sets of nodes in a graph can also be a powerful analysis tool.

There is no single 'right way' to represent a particular set of network data in a graph. Different ways of drawing pictures of network data can emphasise (or obscure) different features of the network structure. It's usually a good idea to play with visualising a network, to experiment and be creative. There are a number of software tools that are available for drawing graphs, and each has certain strengths and limitations. Commonly used package for visualising graphs include KrackPlot (Krackhardt et al., 1994), NetDraw (Borgatti, 2006) and Pajek (Batagelj & Mrvar, 1998).

The Effect of Geography: A Network Visualisation Case Study

An example of the use of visualisation is provided by the following study of the inter-organisational network connecting key tourism organisations in the Gold Coast tourism region of Queensland, Australia. This network was explored using social network analysis techniques (Burt & Minor, 1983; Knoke, 1993; Mitchell, 1969; Scott, 2002; Wassermann & Faust, 1994). The research reported here is part of a larger study jointly funded by Tourism Queensland, a statutory Authority of the Queensland State Government, and the Cooperative Research Centre for Sustainable Tourism. The paper examines differences in the geography, landscape and the type of tourism product in two distinct regions of the Gold Coast – the coastal strip and the rugged hinterland – and relates these to the inter-organisational network structure.

In this study, the Gold Coast has been chosen because of its importance to tourism in Australia as well as its geography and product differentiation. The Gold Coast is Australia's largest holiday destination. In the year to June 2004, 22.5 million visitor nights were spent on the Gold Coast, with domestic visitors accounting for 72% and international visitors accounting for 28% of this total. The Gold Coast is located in southeast Queensland, Australia, just one hour south of Brisbane (Queensland's capital city). Total overnight visitor expenditure in the region was $2.9 billion in 1999, with tourism employing over 36,000 people and accounting for 14.5% of Gross Regional Product in the Gold Coast region (Tourism Queensland 2004).

The study area examined in this paper covers the boundaries of the City of the Gold Coast and parts of the adjacent Beaudesert Shire. Until 1995 the Gold Coast City boundary encompassed a narrow coastal strip of land, in some places only 200 metres wide, from Southport and Surfers Paradise in the north to the NSW border. In 1995, the Gold Coast was amalgamated with the hinterland and northern areas of the Albert Shire Council to form today's City of the Gold Coast.

The City of the Gold Coast is now bounded to the east by the sea, stretches from Beenleigh in the north, then south to the New South Wales border and west to coastal mountains. Beaudesert Shire adjoins the City of the Gold Coast in this mountainous area and extends north into the

Brisbane Valley. Before European settlement, the Gold Coast and hinterland was a natural area of timbered mountains and hills, river valleys, floodplain, salt and freshwater wetlands.

The physical geography of the Gold Coast consists of two regions: the coastal plain and a mountainous 'hinterland' region. The coastal plain is narrow and until recently swampy, with housing activity restricted to the north of the region and the coastal dunes. This led to the development of the Gold Coast as a series of small coastal villages. In the north, Southport and later Surfers Paradise developed as urban areas. Along the beaches to the south, a number of villages grew around life-saving clubs. Later the swamps behind these villages were drained and canal housing estates developed. Today the Gold Coast is a continuous urban strip about 40km long and 10km wide to the base of the mountainous hinterland area. Today the area is an internationally known tourism destination that has substantial high-rise accommodation, both in hotels and apartments, primarily located in the north at Surfers Paradise. Further history of the Gold Coast is provided by Russell and Faulkner (1998) and Prideaux (2004).

The mountainous hinterland of the Gold Coast is composed of the Tamborine plateau to the north, the Darlington and McPherson Ranges, and the Lamington plateau to the south. These areas were originally characterised by large areas of thickly wooded forest, including rainforest remnants. The rainforest areas are today incorporated in a number of small national parks. Early development on the rich soil of the Tamborine Plateau led to small pockets of cattle, dairy and small crop farms. In 1908, under the State Forests and National Parks Act (1906), the first national park was declared at Witches Falls on Tamborine Mountain, 65 km south of Brisbane (Frost, 2004). Later other parks were established in the hinterland. Today this hinterland area is characterised by small rural housing estates, rural farming holdings and semi-rural towns and villages, with the biggest population centre at Tamborine Mountain (population about 5000). Population is denser in the eastern foothills with rural housing estates and 'acreage' properties. This geographic structure and the pattern of economic development of the Gold Coast region has led to two quite distinct areas: a developed coastal beach area, and a green semi-rural region containing a number of national parks. In order to explore the organisation of tourism in these two distinct areas of the Gold Coast, a social network analysis was undertaken.

The fundamental difference between social network analysis and other methods for understanding organisational networks is that social network analysis depends on relational rather than attribute data. The presence and nature of a relationship between actors is the focus, rather than the

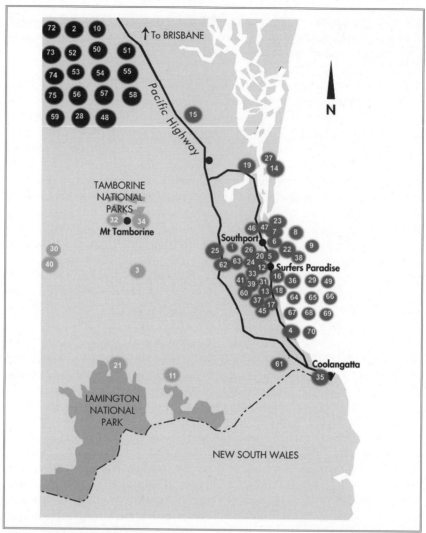

Figure 12.2 The Gold Coast study area and the locations of stakeholders

characteristics of each individual actor. Social network analysis studies may be characterised in terms of the types of social objects being studied, the transactional content, the nature of the links and the resultant structural characteristics examined (Knoke & Kuklinski, 1991). Before collecting data, a network researcher must decide the most relevant type of social organisation to be studied, and the units within that social organisation that will comprise the network nodes. Possible social organisations for study include individuals, aggregates of individuals, organisations, classes and strata, communities and nations.

A network researcher also needs to specify the transactional content of the interaction to be studied. Transactional content refers to what is exchanged when two actors are linked. Different types of transactional content can be distinguished, such as exchange of affect (liking, friendship), exchange of influence or power, exchange of information, and exchange of resources goods or services. Szarka (1990) for example discusses three types of network linkages among small business, based on social interaction, business communication and transactional exchange.

The nature of the linkages between pairs of individuals can be described in several ways, such as intensity or reciprocity. Intensity is the strength of the relation as indicated by the degree to which individuals honour obligations or forego personal costs to carry out obligations (Mitchell, 1969), or by the number of contacts in a unit of time. Reciprocity is the degree to which individuals report the same (or similar) intensities with each other for a content area. In a classic study, Granovetter (1973) discussed the strength of weak ties, highlighting the importance of social relationships in addition to formal business ties.

The structural characteristics studied may be at a number of levels of analysis. The total network involves study of a given set of actors that make up the network and the ways they are linked. Alternatively, an ego network may be studied to define the set of links between one node and all the others to which it is joined. Between these is the study of clusters (groups of closely linked actors), coalitions (temporary alliances of actors who come together for a limited purpose) and 'cliques' (more permanent informal associations that exist for a broader range of purposes).

In developing a social network analysis it is important to ensure the scope of the investigation is delimited by specification of system boundaries (Thatcher, 1998). Determining a boundary for a network study may be done by either focusing on the organisations and their relations (Knoke, 1993), or critical policy events (Pforr, 2002). Thus we may focus on actors sharing a common goal, or use actors located within geographical limits (Laumann *et al.*, 1978: 460). The idea of focusing on actors within a geographical area is related to the study of clusters or industrial districts, as these also have a geographical basis.

Within this boundary the actors to be interviewed must be selected. It may be that all actors within a specified boundary are interviewed but resource limitations usually mean that certain section rules are used. One common method is to distinguish between actors on the basis of their

Table 12.1 Methods used to identify key stakeholders

Method	Actors identified
Positional Analysis	Actors obtained from a formal organizational chart, such as the management positions in companies.
Reputation/Attributional	Actors obtained from selected community members based on judgement as to who is influential or has power.
Decisional Analysis	Actors that participate in making or influencing key decisions.
Interaction Analysis	Actors are asked about their interactions or influence attempts over the period studied.

Source: Knoke & Kuklinski (1991)

degree of influence, and leads to the use of key stakeholders. Various methods and approaches have been used to identify key stakeholders, e.g. position approach, reputation method and decision-method or participation/relational methods (Knoke & Kuklinski, 1991; Thatcher, 1998; Tichy *et al.*, 1979). Each of these approaches has strengths and weaknesses as discussed in Tichy *et al.* (1979) and shown in Table 12.1.

Once a domain and sample have been identified, data collection can proceed. The data sources that may be used for social network analysis are as diverse as in other areas of social research. One recent tourism study used archival analysis to identify respondents, followed by a mail questionnaire (Pforr, 2002). Problems with data collection may be experienced in quantification of relationships and also different types of relationships may be intertwined (Thatcher, 1998). Frequency of exchange of information between organisations is a useful first operationalization of the relationship between organisations. The survey collected information on the relationships among these respondents by asking which other of the key tourist organisations identified they exchanged information with.

Data collection

The initial sampling of organisations for this study was based on identification of key tourism sector stakeholders using the reputation method. Based on initial discussions with staff from Tourism Queensland (the State

Government Tourism Office) and further snowball sampling, 22 key organisations were identified and interviewed. This method allows prioritisation of key stakeholders for contact based on the number of times each was mentioned by other organisations. It also allowed the number of interviews in each region to be reduced to a manageable number. Efforts were made to obtain comments from each of these organisations about each of the other 21 organisations, as well as other organisations considered to be important. As a result, during the study several organisations from outside the region were identified and included but not interviewed due to time constraints. While the study may be seen to use small numbers of respondents, these respondents were perceived as the key stakeholders in the region.

Each organisation was interviewed face to face with the interviews taking around one hour. A written questionnaire was used and in particular respondents were asked a series of semi-structured questions concerning the organisations they had relationships with, the frequency of interaction and the type of interaction. Data was collected using predetermined code frames. A number of open-ended questions were asked during the interview but are not reported here.

Visualising social network data

Analysis of the coded data files was undertaken using the relationship network programs Ucinet 6 (Borgatti *et al.*, 1999) and Pajek. The data collected was visualised using the Spring Embedder technique (Eades, 1984).

The results of the analysis may be illustrated as a series of network diagrams that show the key organisations as nodes and the relationships between them as lines. These network patterns may be subject to both visual and statistical analysis. In this case, visual analysis is used and illustrates the presence of clusters of geographically related organisations on the Gold Coast (see Figure 12.2). This geographical pattern appears related to the pattern of tourism inter-organisational relationships in the region. Figure 12.2 shows a number of nodes representing the organisations interviewed in this study. Each node is labelled with a number for confidentiality purposes. Each of these nodes is connected to one or more other nodes by lines representing reported frequency of communication between those two organisations for tourism purposes. In analysing the results, different types of organisations were coded as different colour nodes: red for organisations on the coastal strip, green for organisations in the hinterland and black for organisations physically located outside the region.

The resultant diagram was produced with no intervention by the analyst. The position of each organisation is derived from the number of links and those of the other organisations to which it has links. The figure produced provides a picture of the social relationship network between organisations interviewed in the region.

Findings

The findings of this study concern the geographical location of the tourism stakeholders interviewed on the Gold Coast in relation to the strength of interaction, as determined by their location on a social network diagram. As shown in Figure 12.2, the geographical locations of the organisations interviewed are concentrated on Surfers Paradise and the coastal strip (grey dots), located in the hinterland (pale grey dots) or located outside the region (black dots). This diagram may be compared with the social network analysis results shown in Figure 12.3, where again the same colours have been used to code nodes.

Figure 12.3 is interpreted as demonstrating the existence of two groupings of organisations on the Gold Coast. The first group consists of tourism operations on the coastal strip. The second group consists of a cluster of

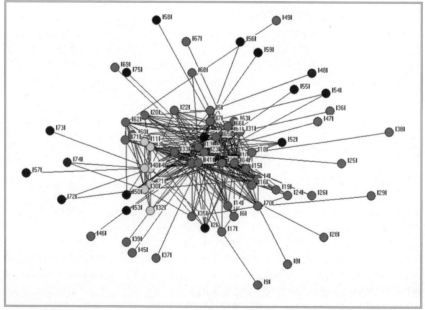

Figure 12.3 A representation of the social network among key tourism stakeholders on the Gold Coast (grey = coastal; pale grey =hinterland; black= outside region)

hinterland organisations who all reported close ties with one another. Such a group is termed a clique. These two groups are linked by the Gold Coast Tourism Bureau (Organisation 33, the Regional Tourism Organisation). The black dots located around the edge of the diagram are primarily government departments based in the state capital of Brisbane. These are public sector organisations primarily involved in planning and projects.

Also of interest is the location of the black dot (51) in the centre of the cluster of grey dots. This indicates the network location of Tourism Queensland, the State Tourism Organisation, outside the region geographically but central to the network relationally.

These findings demonstrate the use of social network analysis techniques in the development of an understanding of the organisation of tourism in a region. Such an understanding is important in management of tourism. The central network locations of the Gold Coast Tourism Bureau and the State Tourism Office indicate their importance to information flows within the Gold Coast. The existence of a clique of organisations in the hinterland however indicates that these organisations are not well integrated into the overall regional network.

The reasons for the existence of this clique are considered here to be partially geographic in origin. The physical separation of the organisations leads to a lack of interaction. Additionally, the separation may also be related to the differences in the type of customers that these operators receive. The relative importance of geography and business is an area for further research.

Conclusion

This chapter and case study have discussed the visualisation of social network analysis techniques. This method is underutilised in tourism and offers some insight. The examination of the empirical structure of tourism networks is a new area for research, and visualisation techniques such as those illustrated here have been found to be very useful for communication of results.

Acknowledgments

This study was funded by Tourism Queensland. The author would like to acknowledge the contribution to this research of Stephen Clark of Tourism Queensland.

Data for this study were collected by Sustainable Tourism Services and Researchworks. Stewart Moore of Sustainable Tourism Services and Nick Parfitt of Researchworks provided valuable input and advice during this study.

Chapter 13
Complex Tourism Networks

Introduction

The recent tourism literature shows a narrow, but important, strand. This considers tourism, and a tourism system, as a complex (sometimes chaotic) adaptive system. The importance of this approach is, according to some scholars, due to the possibility of overcoming the difficulties highlighted in the debate on the disciplinary status of tourism studies. A systemic approach is considered a feasible one to 'reconceptualise' the whole field and to attempt to provide more rigorous theoretical foundations (Farrell & Twining-Ward, 2004; Faulkner & Russell, 1997; McKercher, 1999).

In his seminal paper 'More is different', Phil Anderson states:

> The workings of our mind and bodies, and of all the animate or inanimate matter of which we have any detailed knowledge, are assumed to be controlled by the same set of fundamental laws, which except under certain extreme conditions we feel we know pretty well [but] the ability to reduce everything to simple fundamental laws does not imply the ability to start from those laws and reconstruct the universe [and] at each stage entirely new laws, concepts, and generalizations are necessary, requiring inspiration and creativity to just as great a degree as in the previous one. (Anderson, 1972: 393)

These ideas have contributed to set a new perspective in our view of natural phenomena, a new view which today is known as 'science of complexity' (Waldrop, 1992). Complex systems ideas are amongst the most promising interdisciplinary research themes to have emerged in the last few decades.

This chapter sketches these ideas and briefly presents the tools with which a complex system can be described and analysed. Among these tools, techniques of network analysis techniques are promising for the results that have been achieved so far. An example of the application of these techniques to the tourism field will be given, discussing an investigation of a tourism destination.

Complexity and Complex Systems

The concept of complexity has several meanings in the natural language. Usually it is related to the size and the number of components in a system. Very often, though, by *complex* we merely mean a *complicated* system, i.e. a system composed of a very large number of parts. Such a system, however, can normally be decomposed in sub-elements and understood by analysing each of them; its collective behaviour is the cumulative sum of the individual ones. Any modern machine (a computer, a car, an airplane, a satellite) comprises thousands, or even millions, of single pieces bound together. But, no matter how difficult it is, it is always possible to break up a *complicated* system into separate entities and study them individually, being confident that the final object will be the (linear) composition of them.

On the contrary, a complex system can be understood only by analysing it as a whole, almost independently of the number of parts composing it. There is still no universally accepted definition, nor a rigorous theoretical formalisation, of complexity. Intuitively we may characterise a complex system as 'a system for which it is difficult, if not impossible to reduce the number of parameters or characterising variables without losing its essential global functional properties' (Pavard & Dugdale, 2000). The parts of a complex system interact in a non-linear manner. There are rarely simple cause and effect relationships between elements and a small stimulus may cause a large effect, or no effect at all. The non-linearity of the interactions among the system's parts generates a series of specific properties that characterise its behaviour as complex. A 'simple' object made of only two elements, a double pendulum, a pendulum hanging from another pendulum, is well known to any physics student for its totally unpredictable, chaotic behaviour (under the basic Newtonian laws of motion). A 'simple' school of fish, composed of a few dozen elements, is able to adapt its behaviour to the external conditions without apparent organisation but following a few simple rules regarding local interaction, spacing and velocity (Reynolds, 1987). Generally, a complex system is a mesoscopic structure, composed of a number of interacting elements which is not too low nor too high (Bar-Yam, 1997).

In one special class of complex systems, the *complex adaptive system (CAS)*, interactions among the elements are of a dynamic nature and are influenced by and influence the external environment. In this type of system, the parts 'interact with each other according to sets of rules that require them to examine and respond to each other's behaviour in order to improve their behaviour and thus the behaviour of the system they comprise' (Stacey, 1996: 10).

For a CAS the main characterising features may be summarised as follows:

- *Non-determinism.* It is impossible to anticipate precisely the behaviour of such systems even knowing the function of all its elements. The systems' behaviour depend strongly on the initial conditions; the only predictions that can be made are probabilistic.
- *Presence of feedback cycles (positive or negative).* They influence the overall behaviour of the system.
- *Distributed nature.* Many properties and functions cannot be precisely localised, in many cases there are redundancies and overlaps.
- *Emergence and self-organisation.* The system exhibits *emergent* properties that are not directly identifiable or predictable from the knowledge of the single components. Very often, in a CAS, global structures emerge when some parameter exceeds a critical threshold. In this case a new hierarchical level appears that reduces the complexity. In continuing the evolution, the system increases its complexity up to the next self-organisation process.
- *Limited decomposability.* The dynamic structure is studied as a whole. It is difficult, if not impossible, to study its properties by decomposing it into smaller parts.
- *Self-similarity.* The system will look like itself on a different scale, if magnified or made smaller. Self-similarity is a characteristic evidence of possible internal complex dynamic. The system is at a critical state between chaos and order (self-organised criticality). A self-similar object, described by parameters N and z, has a power-law relationship between them: $N = z^k$. The best known of these laws is the rank-size rule which describes many different objects such as population in cities, word frequencies and incomes. Widely known a Zipf's Law (Zipf, 1949), this rank-size relationship has the general form $P(r) = Kr^{-q}$, where $P(r)$ is the size of the event (population in the case of cities), r is its rank in descending order of size and K is a scaling constant. A power-law means that there is no 'normal' or 'typical' event, and that there is no qualitative difference between the larger and smaller fluctuations.

Examples of complex adaptive systems are considered to be: the patterns of birds in flight or the interactions of various life forms in an ecosystem; the behaviour of consumers in a retail environment; people and groups in a community; the economy; the stock-market; the weather;

earthquakes; traffic jams; the immune system; river networks; zebra stripes; sea-shell patterns; and many others. All these are systems whose behaviour cannot be simply inferred with a linear combination of the behaviours of their elements.

Complexity is a multidisciplinary concept mainly originating in mathematics and physics. It has been applied, sometimes quite extensively, to the world of economics. Indeed, as Saari (1995: 222) states, 'even the simple models from introductory economics can exhibit dynamical behaviour far more complex than anything found in classical physics or biology'.

A great number of studies, both from a theoretical and empirical point of view, have been produced in recent years, concerning many aspects of this world: macro economies, markets, stock exchanges, down to single sectors or even single firms of any size (Arthur *et al.*, 1997; Mantegna & Stanley, 2000; Mirowski, 1989; Sterman, 2000). As a 'practical' outcome of these studies, for example, companies and organisations are recommended to modify their established management practices if they want to survive and develop in the modern world, and to adopt new management theories:

> Because businesses are complex adaptive systems, nested in larger complex adaptive systems (the economy), managers should always expect surprises, no matter how carefully they plan, or how simple the goal. Indeed, they should not even attempt to plan too precisely, because inevitably a linear approach will fail in some respect or other as the business environment constantly changes. (Lewin, 1999: 202)

Tourism systems, like other economic activities, can be thought of as complex systems. Complexity theory offers the hope of being able to understand, for example, how disasters or turbulent changes may influence the sector, or why, after major global or local crises the tourism sector is able to show a rapid and almost unexpected recovery (UNWTO, 2002). The theoretical work in this field is still in its infancy. Just a handful of researchers have begun to consider the complex systems approach as a more effective framework for the understanding of the many and different phenomena (Farrell & Twining-Ward, 2004; Faulkner, 2000; McKercher, 1999). Their analysis, so far, has been mainly performed on a qualitative basis. More recently, however, a quantitative assessment of many of the complexity characteristics of a tourism system has been provided (Baggio, 2007b).

As many authors generally note, the great majority of the approaches to the study of tourism systems are industry-centred and take little care of elements from the external environment that influence tourism outcomes.

Moreover, a basic assumption is that plans and controls can be more or less easily put in place and that, with sufficient information, know-how, policy development and regulations, good outcomes can be achieved. On the other hand, unexpected events such as terrorism, climate changes and diseases, or more simply, the behaviour of tourists, can generate, as has been experienced in recent years, very diverse effects. At times insignificant episodes are able to induce catastrophic consequences. One of the main conclusions of complex system thinking is that different behaviours strongly influence a system, and its performance is largely unpredictable. Therefore, all the basic traditional methods for forecasting and managing a tourism system need a deep revision with a shift towards an adaptive attitude, rather than a rigid deterministic style. Exploring and simulating alternative possibilities, implementing one or more of them, monitoring the outcomes, testing the predictions and learning which one most effectively achieves management objectives has already provided encouraging results (Agostinho & Teixeira de Castro, 2003; Caffyn & Jobbins, 2003).

Complexity sciences also embrace, according to some scholars, chaos theory (Lewin, 1999). In essence, a system may be considered as evolving from a completely ordered phase to one in which its behaviour is so strongly dependent on very small variations of the initial conditions as to appear to be almost unpredictable: a 'chaotic' phase. In this, still governed by deterministic laws, the system may tend to certain specific configurations. These, the *attractors*, and the regions close to them (their *basins of attraction*), can be fixed equilibrium points, orbits or more complicated patterns. It also possible to have a system that never returns to the same place (in these cases we speak of *strange attractors*). The region at the boundary of these phases, the one which has been called the 'edge of chaos', is a region of complexity (Crutchfield & Young, 1990; Waldrop, 1992). Chaos theory studies non-linear effects on deterministic systems, while complexity theory studies definite patterns on non-deterministic systems (Gleick, 1987; Kauffman, 1995).

The toolbox available to study a complex system has become, in recent years, very full. Many of the methods and the techniques originate from the work of 19th century scientists, but modern computational facilities have made them amenable to calculation. In their review, Amaral and Ottino (2004) identify three main classes of analysis tools, each belonging to an area well known to physicists and mathematicians: nonlinear dynamics, statistical physics and network theory.

Statistical physics (or statistical mechanics) is one of the fundamental fields of physics. It uses statistical methods for addressing physical problems. A wide variety of issues, with an inherently stochastic nature, are treated in such a way. It provides a framework for relating the microscopic properties of individual atoms and molecules to the macroscopic ones of materials observed in everyday life. Traditionally it has provided the background for the study of phenomena such as the phase transitions (and the criticalities involved in these) or the study of systems and their exchanges of energy with the external environment. More recently it introduced the idea of discrete models, cellular automata and agent-based models (Tesfatsion & Judd, 2006; Wolfram, 2002; Shalizi, 2006; see also the discussion in Chapter 7).

Two important concepts are founded in statistical physics: universality and scaling (Amaral & Ottino, 2004). Universality is the idea that general properties, exhibited by many systems, are independent of the specific form of the interactions among their constituents, suggesting that findings in one type of system may directly translate into the understanding of many others. Scaling laws govern the variation of some distinctive parameter of a system with respect to its size. Its mathematical expression for complex and chaotic systems involves a power-law, which is considered a characteristic signature of self-similarity.

Most complex systems can be described as networks of interacting elements. In many cases these interactions lead to global behaviours that are not observable at the level of the single elements and that share the characteristics of emergence typical of a complex system. Moreover, the collective properties of dynamic systems composed of a large number of interconnected parts are strongly influenced by the topology of the connecting network (Barabási, 2002; Buchanan, 2002). This approach is the one we shall follow in the remainder of this chapter.

The Contribution of Network Analysis to Tourism Studies

Although network analysis methods are quite 'old' and tourism is a network business, these techniques have not been widely applied to the study of the tourism sector. Some recent contributions, however, show the usefulness and the effectiveness of this approach.

The networking of single individuals and the diversity of their ties plays a significant role in the decision routines in the tourism industry. Furthermore, it has been noticed that there are differences in the use of efficient ties, in terms of strong and weak ties. They change depending on specific tasks and successful actors are able optimise their composition (Pesämaa & Skurla, 2003). This is a reconfirmation of results found in completely different disciplines such as biochemistry (Csermely, 2004). These studies validate the importance of a balanced and heterogeneous composition of links as a factor fostering the success of a tourism network (Pavlovich & Kearins, 2004).

Pforr (2001, 2002, 2006a) shows the usefulness of a network approach in describing, analysing, and explaining the dynamics of the tourism policy realm. Furthermore, measuring link characteristics leads to effectively examining the role of different actors in the network. Thus, it is possible to see how actors deemed as relevant for strategic planning processes turned out to be more or less significant for the real plan formulation processes.

The formal network approach, besides highlighting the differences in the organisation of tourism activities in different destinations, proves useful in stressing the necessity of collaboration and cooperation, typically lacking in this sector (Bramwell & Lane, 2000). The emphasis on the formation of a value creation system through a balanced set of relationships substantiates that, and offers guidance to policy makers and management organisations (Scott & Cooper, 2006, private communication).

Following this line of research, Dredge (2004, 2006a; Dredge & Jenkins, 2003) suggests that network theory provides a tool for understanding the relationships between government, business and civil society, and how these may influence and facilitate collaboration and cooperation among the different actors. Local tourism planning, it is found, cannot be based on a simple conjecture of agreement and willingness to collaborate. Policy implementations favouring this state may prove effective if they are achieved through adaptive incremental steps. It is not only political and strategic tourism issues that benefit from network analysis, but also more 'practical' problems. Shih (2006) shows, in fact, how to use these methods to revise the organisation of tourist facilities and services in particular destinations by mapping and measuring the structural characteristics of routes taken by tourists in multi-destination trips.

In conclusion, sound methods for the analysis of networks have a great importance for the study of a tourism destination, not only as a fascinating intellectual problem, but also as a means to improve abilities and capabilities in understanding the functioning mechanisms of a tourism destination in order to manage it effectively and efficiently.

A Methodological Approach

The analytic framework described in the previous chapters provides the basis for understanding how a tourism district can be represented and analysed. This is important in determining what data is relevant and needs to be collected (Yin, 1994). The unit of analysis considered here is a tourism destination composed of a collection of organisations (public or private) and their common relations. The destination is therefore a network whose actors are the single organisations and whose links are the connections established among them.

The most important issue for the overall validity of an investigation in the field of complex social networks is the collection of data. The problem is well known since the seminal work by Ove Frank on the statistical inference in graphs (Frank, 1971). Social network analysis literature has dealt quite extensively with this issue and with the reliability of the different methods proposed (Marsden, 1990). The analysis of a network relies on data collection procedures that may be difficult to execute, or lead to incomplete or unreliable outcomes. The collection is typically based on surveys and interviews with informants, with a number of different techniques that aim to highlight the connections among the different actors.

As long as we think that a network is a number of nodes connected by a number of links whose distribution is more or less random (as in the Erdõs-Rényi model), the issues of sampling a number of these elements and analysing them in order to derive the whole network characteristics are well known and studied. Classical statistical methods allow a wide range of possibilities to estimate, infer, generalise and model some characteristics of a set of objects (a population) by studying the same characteristic in a subset of elements (a sample). These well-established methods also give innumerable ways of assessing validity, confidence and reliability of measurements and conclusions (Cochran, 1977; Gentle et al., 2004; Langley, 1971; Shao, 1999; Sheskin, 2000).

With our present knowledge of the different network topologies, however, the situation is quite different. Ignoring part of the elements, or lacking part of the data, can have a direct influence on the estimation of the main network parameters. Simulation studies show that clustering and assortativity coefficients are overestimated when both actors and relations are omitted, and underestimated when ties are missing or incorrectly reported in surveying the network components (Kossinets, 2006). On the other hand, when the links are correctly identified, some centrality measures (in-degree and simple eigenvector) are found to be relatively insensitive to the sampled number of vertices in the network, so a reliable estimate is possible even with 50% of nodes missing (Costenbader & Valente, 2003).

The effects of an incomplete sampling of network components (both nodes and links) on the structural characteristics can be generally assessed by simulating this process. We may start with a network of a certain size and remove some part of the nodes or the links, recalculate the parameters we are interested in and assess the difference. This procedure is based on the results obtained on the analysis of the robustness of complex networks connectivity with respect to random failures or targeted attacks (Albert *et al.*, 2000; Crucitti *et al.*, 2004).

For example, Figure 13.1 shows that the overall connectivity of a scale-free network is quite robust under random node removals (failures), while it is easily lost under coordinated attacks that affect just a few highly connected nodes. On the other hand, no significant differences are found if the network has a random distribution of links. In general, it is clear that, depending on the basic topology, an incomplete network may show properties that have different characteristics from those possessed by the complete system.

The relationship between the network parameters measured on a partial sample, and the complete one, can be inferred by simulating a sampling process and by calculating the difference effects. With this approach, a number of recent studies have provided insights into these issues (Han *et al.*, 2005; Lee *et al.*, 2006; Rafiei & Curial, 2005; Smith *et al.*, 2003; Stumpf *et al.*, 2005a; Stumpf & Wiuf, 2005; Stumpf *et al.*, 2005b).

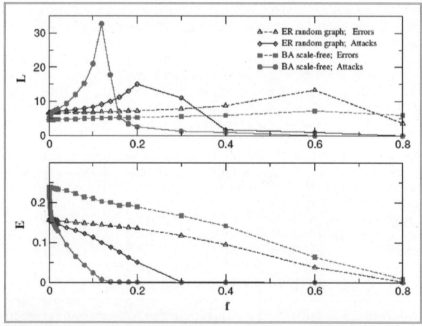

Figure 13.1 Effects of random removals (errors) and targeted attacks (attacks) for random (ER) and scale-free (BA) networks (*f* is the fraction removed) on average path length (L) and efficiency (E) of the system (after: Boccaletti *et al.*, 2006)

Table 13.1 Effects of sampling methods on SF network properties

	Degree exponent	Average path length	Clustering coefficient	Betweenness exponent	Assort -ativity
Node ⇓	⇑	⇑(⇓)	⇕	⇑	⇔
Link ⇓	⇑	⇑(⇓)	⇓	⇑	⇔

Table 13.1 shows the behaviour of the main network parameters, in the case of a scale-free network, for the sampling of the different network elements. It must be noted that the average path length may have a stronger dependency on the average degree, decreasing with the increase in degree; this effect may overcome the one shown in the table for peculiar cases.

If the network has an Erdõs-Rényi (or, generally, exponential) topology, the dependency is different (as discussed above). For example, a random sample preserves the shape of the degree distribution while modifying the average degree value (Figure 13.2).

These results provide general criteria for sampling methods when some specific parameter is investigated, and allow a better evaluation of the analysis results. By estimating the sampled fraction, therefore, it is possible, at least, to determine bounds for the main network parameters.

A Comparison Between Tourism Destinations

Let us now consider two different tourism destinations: Fiji Islands and the island of Elba, Italy (Baggio, 2007a; Baggio *et al.*, 2007a; Baggio *et al.*, 2007b). The destinations share many similarities:

- In both cases they offer 'sun and sand' tourism experiences.
- They both receive tourist arrivals of around 500,000 visitors per year who spend around 3 million visitor nights.
- Each offers accommodation capacity of the same order of magnitude (some ten thousands bed places).
- Both economies are highly dependent on the expenditure generated by tourism activities.

The analysis considers the network formed by the core tourism stake-holders of both destinations (accommodation, intermediaries, transport, regulation bodies and services) via their websites' connections. We base this investigation upon the assumption that the connections among the websites (hyperlinks) may be considered not simply as a technological manifestation but also as a reflection of social processes. The structure of hyperlinks forms patterns based on the designs and aspirations of the individuals or

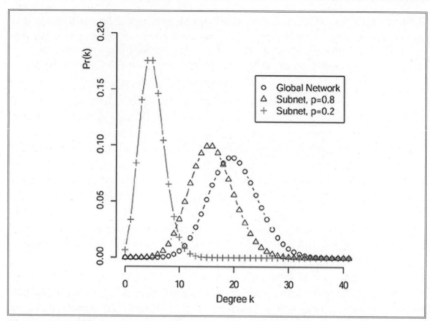

Figure 13.2 Effects of the sampling fraction (p) on the degree distribution of an exponential (ER) network (after Stumpf & Wiuf, 2005)

organisations who own websites. A growing literature suggests that these networks reflect offline connections among social actors and support specific social or communicative functions (Jackson, 1997; Park, 2003; Park & Thelwall, 2003). The layout of a network of websites for a tourism destination can thus be seen as a reflection of characteristics of the structure of the social network from which it originates. This relationship between cyberspace and the physical world is two-way: on one side, the online linkages represent and complement social relations in the offline world; on the other side, offline interactions can influence how online relationships are established and developed (Birnie & Horvath, 2002; Wellman, 2001).

The main structural characteristics are measured, both from a static and a dynamic point of view. The websites have been analysed considering them as the nodes of a complex network. The elements of the network have been identified by using official lists provided by the Elba Tourism Board and Fiji Visitors Bureau. All links are considered of directed nature. The analysis has been performed with Pajek (de Nooy *et al.*, 2005) and a set of programs written by one the authors using the Matlab (MATLAB, 2004) development environment. The sizes of the networks examined are 468 elements (websites) for Elba and 492 for Fiji.

The graphical representation of the two networks is given in Figure 13.3 (Elba) and Figure 13.4 (Fiji). As can be seen, both networks exhibit an identifiable structure. This is clearer if we compare them with the graph obtained by generating a network of a comparable size (500 nodes) and link density (2%), whose links are purely randomly distributed (see Figure 13.5).

The numeric values calculated for the main characteristic parameters of the two networks are given in Table 13.2. These characteristics indicate that both networks are rather sparse, showing very low densities and high proportions of totally unconnected websites. Diameters and average path lengths are almost in line with those exhibited by similar networks (Albert & Barabási, 2002; Dorogovtsev & Mendes, 2002). The clustering coefficient and the local and global efficiency of the graphs are very low as well and considerably lower than those found for similar systems. The local efficiency value confirms the poor clustering of the network. Finally, it is interesting to note that for both networks the assortative mixing coefficient is low and, more importantly, negative. This is the opposite of what is commonly found for social networks in which, typically, the most connected nodes tend to link nodes with similar degrees (Newman, 2003).

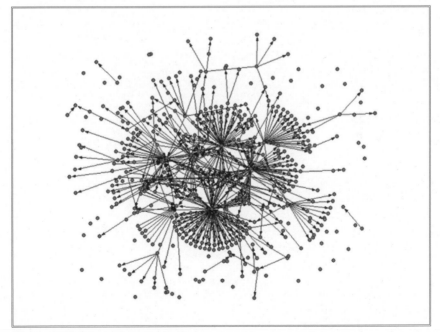

Figure 13.3 Elba network graph

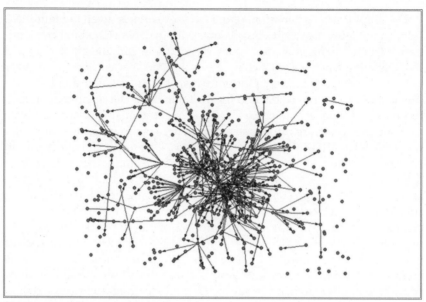

Figure 13.4 Fiji network graph

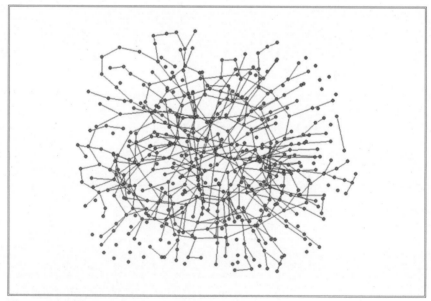

Figure 13.5 The graph of a random network

Table 13.2 Main characteristics of Fiji and Elba networks

		Fiji	*Elba*
Size (# of nodes)		492	468
Link density (δ)		0.0016	0.0023
nodes with no connections		35%	21%
Average path length (L)		2.9	4.5
Diameter (D)		6	11
Clustering coefficient (C)		0.024	0.003
Efficiency			
	local	0.0275	0.0145
	global	0.0710	0.1698
Assortative mixing coefficient		-0.137 ± 0.102	-0.101 ± 0.094

The cumulative degree distributions are shown in Figure 13.6 (Elba) and Figure 13.7 (Fiji); (k_{in} is the in-degree distribution and k_{out} the out-degree distribution (the distributions of the number of links incoming to a node or outgoing from it).

For the Elba network, both distributions follow a power-law whose exponents are: out-degree: $g_{out} = 1.89$ and in-degree: $g_{in} = 2.96$. In the Fiji network a pure power-law behaviour seems to be attributable only to the in-degree distribution (exponent is: $g_{in} = 2.91$). The out-degree distribution shows a clear cutoff at high k values. The best fit for this distribution is therefore a function of the form: $P(k) \sim k^g \, exp(-k/k_c)$, with $g_{out} = 1.4$ (the central part of the distribution scales as a power-law) and $k_c = 15$.

Both in-degree exponents are quite high (Albert & Barabási, 2002; Dorogovtsev & Mendes, 2002), indicating a great concentration of the networks. The values found are consistent with the preferential attachment growing mechanism suggested by Albert and Barabási (2002). Out-degree exponents are, instead, rather low, sign of a relatively flat and spread distribution of the links. A spectral analysis confirms the main topological characteristics of the networks, also acknowledging their scarce connectedness.

In summary:

- Both networks show a scale-free topology (power-law behaviour of the degree distributions) which is consistent with the one generally ascribed to many artificial and natural complex networks.

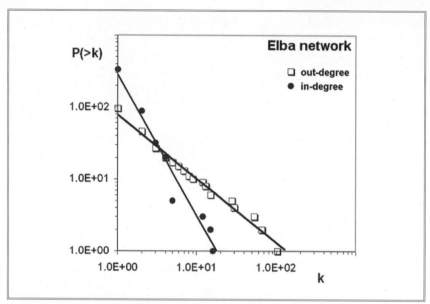

Figure 13.6 In-degree and out-degree cumulative distributions for the Elba network

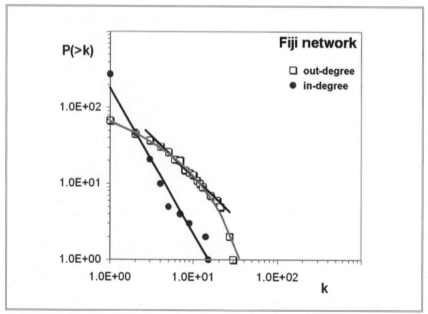

Figure 13.7 In-degree and out-degree cumulative distributions for the Fiji network

- The general connectivity is, however, very low (link density) with large proportions of disconnected elements.
- Clustering is quite limited, as is the efficiency both at a local and at a global level.
- A very small and negative correlation among the degrees of the nodes has been found (assortativity coefficient), i.e. nodes with high degree tend to connect with nodes with low degree. This behaviour is the opposite of that typically found for social networks.

As discussed above, we may also assume that the network of destination websites represents more than an artificial technological network, and the web space of a tourism destination is a representation of the underlying economic and social network. In this context, our analysis can provide interesting insights. The clustering coefficient and the assortativity index may then be used as quantitative assessments of the degree of collaboration or cooperation among the tourism destination's stakeholders. In this case, the clustering coefficient can be thought of as a *static* measurement of cooperation, and representing the formation of cohesive communities inside the destination, while the assortativity measure can be interpreted as representing the tendency to form such communities.

Under this assumption, the general low connectivity and low clustering characteristics of both networks are a clear indication of very limited degree of collaboration or cooperation among the stakeholders. The negativity of the assortative mixing coefficient also reinforces this reading. A confirmation of this interpretation comes from previous studies on Elba (Pechlaner *et al.*, 2003; Tallinucci & Testa, 2006) where it has been argued that a low propensity to connect to the external world exists. The reason is the strongly independent way in which small family-run enterprises (the vast majority of the tourism businesses on the island) are conducted. In the case of Fiji, the structure of Fiji's tourism industry is mainly based on 'all inclusive' resorts that are self-sufficient and have little collaboration with other organisations besides the very basic supply chain relationships.

As seen above (Table 13.2), some differences in the values of the network parameters exist. Fiji values show a lower connectivity, a higher degree of disconnected elements, lower efficiency and lower assortativity. Its relatively higher clustering coefficient can be explained by the smaller size of the 'giant component', the largest connected component in the network.

Let us consider now the degree distributions (Figure 13.6 and Figure 13.7). The out-degree distribution of the Fiji network exhibits a marked cutoff at high k. This characteristic, for a complex network, is usually interpreted as the result of some kind of constraint on the evolution (growth) process of the network. Constraints can be in the form of cost limitations in forming connections, aging of nodes who stop creating links after a certain period of time, spatial confinements of the network or finite lifetime (Boccaletti *et al.*, 2006; Dorogovtsev & Mendes, 2003; Dorogovtsev *et al.*, 2000; Jin *et al.*, 2001; Rodgers & Darby-Dowman, 2001).

Before being able to give a meaning to these characteristics, we need to consider the tourism characteristics of both destinations. Elba is considered a 'mature' tourism destination (Tallinucci & Testa, 2006). It has a long history and has gone through a number of different expansion and reorganisation cycles. The stakeholders are mainly small and medium companies and there are a number of associations and consortia who try to overcome the 'independence' of the companies by implementing different kinds of collaboration programmes. Geographic, economic and political factors have not favoured a full development of tourism in Fiji (Harrison, 2004). The destination is divided into a number of different geographical locations such as the Coral Coast, Mamanucas and Yusawas and, as noted above, the supply structure is highly fragmented. Only recently are central tourism policy organisations designing and implementing coordinated development plans.

With this scenario, and if we accept the idea that a tourism destination has some kind of evolutionary path (Butler, 2005a, 2005b), we may legitimately say that Fiji is at an earlier stage of development, as a tourism destination, than Elba. In an early stage of development, tourism organisations exist, but they have not yet connected to others. This happens because they do not feel the necessity or because they have not yet realised the existence of other companies. Larger organisations or associations, generally responsible for the higher degrees, still have to establish a link with the newer nodes in the network. In other words, there seems to exist a limitation in (some of) the nodes' ability to process information about all the other nodes of the network. This filtering of information is able to generate (Mossa *et al.*, 2002) the exponential truncation found in the degree distribution of Fiji websites. Moreover, as these authors say, concluding their paper:

> In the context of network growth, the impossibility of knowing the degrees of all the nodes comprising the network due to the filtering process – and, hence, the inability to make the optimal, rational, choice – is not altogether unlike the "bounded rationality" concept of Simon (1997). Remarkably, it appears that, for the description of WWW growth, the preferential attachment mechanism, originally proposed by Simon (1955), must be modified along the lines of another concept also introduced by him – bounded rationality (1997). (Mossa *et al.*, 2002)

This kind of relationship between the modifications in the network topology and the evolution of a destination is similar to the one described, if only at a qualitative level, by Pavlovich and Kearins (2004) in their analysis of Waitomo (NZ).

Conclusion

The techniques and methods of analysis of complex networks have developed significantly. An increasing number of research works have studied a great variety of theoretical and empirical aspects of networks of many types. Tourism systems, in spite of their recognised 'networked' characteristics, have been almost absent from these investigations. Only a handful of papers have employed basic social network analysis methods, and generally only at a qualitative level, in this sector.

We have discussed the methodological issues involved, and mainly the ones concerning the collection of basic data. As a practical example we have considered two tourism destinations and, by examining their web spaces, have provided an assessment of the main characteristics of the two networks. With the hypothesis that the technological system represents the economic and social group that originates it, we have derived a series of characteristics of the structure and the cohesiveness of the system, emphasising the aspects connected with the issues of collaboration and cooperation. Moreover, by assuming the existence of an evolutionary growth for a tourism destination, we have connected the basic topological characteristics of the networks with such evolution.

Clearly, the exploratory nature of the analysis and the limited number of examples examined puts a limit on the generalisation of our results. However, even such limited results show, in our opinion, the validity of the methods used. The possibility of giving quantitative assessment to structural characteristics of the network of stakeholders in a destination may prove extremely useful for the organisations responsible for the management of the system. Until recently only qualitative measurements of network characteristics were possible, with all the reliability limitations such investigations have. The approach used here provides a quantitative confirmation of the level of collaborative phenomena, and may better help in directing planning activities. Moreover, the usage of computer simulation models, which may be easily implemented once the basic structure of a system is known, can provide useful tools to study different possible optimisation schemes and allow the building of more reliable development scenarios (see Chapter 14).

Technological Tourism Networks and Network Simulation

Introduction

Information and communication technology (ICT) has profound implications for the tourism industry and is being used extensively in a great variety of functions. Contemporary ICTs have radically altered the way in which information is conveyed throughout the industry (B-to-B) and to customers (B-to-C) with the effect of bridging the distance between all those involved in this market. The uses of ICTs are driven by the interaction between users' demand requests, and the rapid expansion and sophistication of new products and services (both hardware and software) offer new ways of delivering old products and services. In this chapter we discuss the role of ICTs in the tourism world. The general structure of the Internet will then be examined. Finally, the examination of a case study will show how network analysis methods can be used in this context.

Among the modern ICTs, the Internet has become, in less than two decades, the most significant development in communications since the invention of the printing press. The Net, as it is called, has revolutionised deeply the way in which we communicate, conduct business or research, and study; in a word, the way we live. The number of users connecting to the Internet worldwide is estimated at almost one billion, almost 20% of world's population, and in some areas the percentage reaches 70% (IWSTAT, 2006). Its rate of diffusion, still growing, has been much faster than radio in the 1920s, television in the 1950s or mobile phones in the 1980s (Odlyzko, 2000).

The Internet and its most renowned service, the World Wide Web (WWW), provide features that are especially relevant to the marketing of tourism. Travel is considered to be an experiential practice; travellers are not simply buying packages, stays, seats, or food and beverages, they are fulfilling fantasies (Archdale, 1995). In this regard, the quality and quantity of information on the products, and the way of presenting them to potential customers, plays a crucial role in directing their choices (Buhalis, 1998). ICTs, as a consequence, have an essential function in this process. It is no surprise that tourism is one the most important application areas on the WWW: 'In few other areas of activity are the generation, gathering, processing, application and communication of information as important for day-to-day operations as they are for the travel and tourism industry' (Poon, 1993: 154). In today's tourism world, using ICTs and the Internet is no longer a distinctive characteristic by itself, and only effective and efficient use of it can help in obtaining a competitive advantage. But when the right technologies are available and are correctly applied, organisational benefits and growth become quite visible (Buhalis, 1998; Ragowsky *et al.*, 2000; Werthner & Klein, 1999).

A great deal of study, research and analysis has been carried out in recent years on the Internet, its usage and importance in different fields and for tourism in particular. In particular its network characteristics have been investigated from many different perspectives, both from a physical point of view (connectivity among computers, telecommunication lines and other devices forming the network) and a logical one (the contents presented on network: websites, applications, services). In addition, the Internet has stimulated the field of complex network analysis. The wide availability of data on millions of interconnected machines or billions of websites and web pages has provided the material from which to develop models and theories on the structure and the behaviour of complex networks. The three seminal papers by Faloutsos *et al.* (1999), Barabási and Albert (1999) and Watts and Strogatz (1998), which stimulated the application of complexity theory to networks and led to the our present understanding of the concepts of scale-free and small world networks (see introduction and history of network theory chapters in this book), all use some subset of the Net as a primary source of data. Without it we would still work with the Erdös and Rényi (1959) random network model and the limited possibilities that this simplified view has in explaining a wide number of phenomena (such as diffusion, critical transitions or robustness).

ICTs, Internet and Tourism Research

As stated above, a great number of academic researchers have examined ICTs and tourism. All scholarly journals involved in the field publish works on this topic (see for example the reviews by Frew, 2000 and O'Connor & Murphy, 2004). Moreover a dedicated journal exists: *Information Technology and Tourism* (*IT&T*), and an international association IFITT (International Federation for IT & Travel and Tourism) holds an annual, well-attended international conference: ENTER. The topics investigated by academics cover all aspects of the matter: the behaviour of consumers using electronic media, the effectiveness of a website as a promotional tool, the benefits of full integration between different types of media or between internal computerised information systems, and the world of the Net. A scan of the last five years worth of papers from the ENTER conference and the *IT&T* journal (more than 350 papers) shows that, within the general theme of information technology, more than 80% of the works have the Internet or some part of it (web, email, extranet, intranet, wireless access, etc.) as a major argument.

In scanning this literature we may note a highly fragmented research base and diversification of models, methodologies and approaches applied to these common issues. Despite this, there is agreement on some basic findings. For example, let us consider the evaluation of the quality characteristics of a website. This is important for at least two reasons: it can provide managers with key information useful to maximise the returns (tangible or intangible), and it can help in studying the behaviour of the users and their reactions to the contents and services offered online. This, for a tourism organisation, is of fundamental importance as it has also a direct influence over the image of the organisation on the market (Baggio, 2005; Detlor *et al.*, 2003; Steinfield *et al.*, 2001). A literature scan shows that each single author has their own methodology to assess the acceptance of a website, even if the conclusions of most of these works converge to a common series of suggestions and recommendations on the necessity of providing good and updated contents, useful and usable interactive services and pleasant but simple graphic layouts (Antonioli Corigliano & Baggio, 2006). This situation is not uncommon in the scholarly studies in tourism, and it has been discussed several times (Echtner & Jamal, 1997; Farrell & Twining-Ward, 2004; Leiper, 2000; Tribe, 1997), with a particular emphasis on the necessity of a more rigorous theoretical approach to this field of studies.

A second interesting conclusion of the scrutiny of ICT and tourism literature is that although a network of networks (the Internet) is a largely dominant subject, very little has been used so far of the methods and the techniques developed in the field of network science. The rest of this chapter discusses the application of network science to tourism and ICT, and the implications that can be derived from such a study.

The Web as a Complex Network: Basic Models

The Internet can be studied at two different levels. The first is the one consisting of the *physical* machines, the undirected network of interconnected vertices composed of hosts (end-user computers), servers (providing a network service to other computers), and routers (that drive the data traffic across the network). Servers, routers and IP (internet protocol) networks managed or owned by one entity (organisation, company etc.) are grouped in autonomous systems (Hawkinson & Bates, 1996). An autonomous system (AS) may be roughly identified as an Internet domain. The second one, which we could define as a *logical* level, is the ensemble of contents (information, data, services) provided on the Net through the service known as the World Wide Web (WWW or Web).

The WWW is, without any doubt, the most important and significant phenomenon of the last few years. Billions of pages, sites and databases are offered on the Net and allow us to perform a wide variety of functions (e.g.: search for information, play games, sell products or communicate with other people). The impact on almost all the forms of our social, economical and personal life has been enormous. The Web remains essentially uncontrolled by any governing body: any organisation, institution or individual can create a website of any size, with relatively little effort. This growth, without clear rules or centralised control, produces a huge, self-organising, dynamic complex system.

In network science terms, the WWW can be described in terms of a large directed graph whose vertices are the html documents and whose edges are the hyperlinks connecting one document to another. As discussed in earlier chapters, in a complex network the topological structure determines its connectivity characteristics, such as the efficiency in data transfer or response times or capability to connect any two elements of the network. The study of the Web as a graph is not only intriguing per se, but it is also of great importance in providing 'practical' answers to significant problems. For example, one issue is *visibility*, or the problem of finding functional algorithms for crawling (Deo & Gupta, 2001), searching and discovering websites or communities (Gibson *et al.*, 1998; Newman & Girvan, 2004). A casual user may be able to find one of the several billion existing websites by accessing a search engine. Despite their name, these consist of *static* databases fed by specialised software applications (robots or crawlers) whose task is to scan the whole Web looking for new or updated contents and adding them to the database. Given the number of *objects* on the Net, a sequential search of all the possible IP addresses is obviously impossible. Therefore any efficient method capable of shortening this process while ensuring effectiveness in the results is of paramount importance for both the organisation providing the search engine functionalities and the users accessing them. Moreover, visibility is an economic issue, as there is value in ensuring that tracking down a website is as easy as possible, and therefore connections (hyperlinks) can be seen as having a pseudo-monetary value (Walker, 2002).

Many metrics have been proposed to measure and characterise webpage–hyperlink networks (Dhyani *et al.*, 2002) and most of them are a direct derivation of those used for analysing any kind of network. Empirical studies (Albert & Barabási, 2002; Boccaletti *et al.*, 2006; Pastor-Satorras & Vespignani, 2004) show that, at both levels (*physical* and *logical*), the Internet exhibits the features of a complex, scale-free network with distinctive small world properties. It has a good level of modularity (relatively high clustering coefficient) and high compactness (short average distances and diameter). Table 14.1 shows some typical values for these characteristics.

Table 14.1 General network characteristics of the Internet

Network	mean degree	average path	diameter	clustering coefficient	Degree distribution exponent		
					in	out	undirected
AS	4.1	3.6	10	0.24			2.4
Routers	2.6	3.5		0.03			2.2
WWW	7.5	16	7	0.11	2.1	2.7	

Note: values are weighted averages of those cited in literature.

Before discussing the characteristics of the realm of the network of tourism-related websites, it may be worth examining the general features of the whole web space (the whole WWW). Intuitively, as integral part of this larger world, the tourism sub-Web (if we can define it as such) will share most of the peculiarities of it.

Different general models for the structure of the Web have been proposed. The most frequently used model is based on the research by Broder *et al.* (2000), Dill *et al.* (2002), and Flake *et al.* (2002). This is known as the 'bow-tie' model, for the shape with which it has been depicted by the original proponents (see Figure 14.1). It considers the Web as a self-organising, self-similar structure divided into three main components:

- A core of strongly connected nodes (SCC), accounting for almost 28% of the web pages in the sample studied, all joined with bidirectional links.
- A set of pages (IN) mainly connected in a monodirectional way to the SCC. The pages in this component (almost 21% of the sample) have

outgoing links that reach the SCC, but they are virtually unreachable by other parts of the web.

- A similar sized (21%) set of pages (OUT) reachable by the pages contained in SCC, but whose links are mainly inward bound; i.e. there is always a path from SCC to OUT pages, there is no direct connection from OUT to SCC or IN.

The picture is completed by two more sets: TENDRILS composed by pages (again 21%) providing paths from IN pages or to OUT pages without passing through the SCC elements. The TENDRILS contain pages that cannot reach the SCC, and cannot be reached from the SCC. It is possible for a TENDRIL hanging off from IN to be hooked into a TENDRIL leading into OUT, forming a TUBE – a passage from a portion of IN to a portion of OUT without touching SCC. To complete the model, there are a certain number of

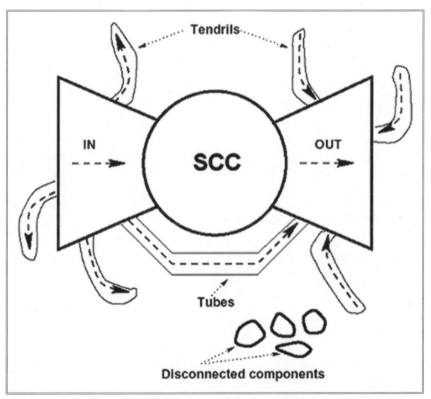

Figure 14.1 The bow-tie, a general model for the Web space (after Broder *et al.*, 2000)

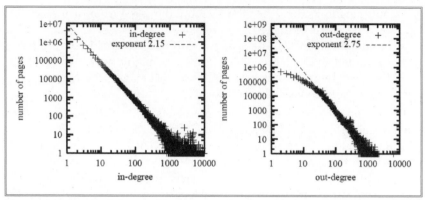

Figure 14.2 Typical degree distributions for the web (after Guillaume & Latapy, 2002)

pages disconnected from the other main components (DCC). The general picture, a bow-tie like graph as said, is shown in Figure 14.1.

Based on the 'bow tie' model we can see that the supposed high global connectivity of the Web appears, in fact, to be not so high. Only in 24% of the cases it is possible to find a path between any two nodes chosen at random (Broder *et al.*, 2000). The average distance (average shortest path) between nodes is ~16, and it decreases to 6.8 if we consider the undirected connections in SCC.

The degree distribution, on average, follows a power-law. The values of the exponent γ are: $\gamma = 2.1$ for the in-degree distribution (incoming links to a page) and $\gamma = 2.72$ for the out-degree distribution (links outgoing from a page). These values, or values very close to these, have been confirmed by several different empirical studies, that confirm also the value for the web diameter $D \sim 7$, and the average degree of a node in the Web $<k> = 7$ (Pastor-Satorras & Vespignani, 2004).

The web network also shows self-similar characteristics. Dill *et al.* (2002) have shown that the whole web can be thought of as a modular structure in which different parts (called thematically unified clusters, TUC), each one exhibiting a bow-tie structure, combine to form the larger web space system (Figure 14.3). The connected components of these TUCs are strongly integrated into the general SCC of the web. The authors also claim that the degree of integration of the SCCs into the wider web network can be taken as an indication of how well established the community is.

The topological characteristics of the web networks, however, are not completely uniform. As has been shown (Pennock *et al.*, 2002), in fact,

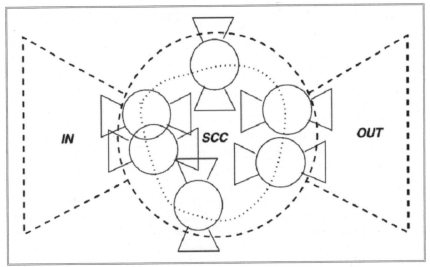

Figure 14.3 Self-similarity of the web structure

Table 14.2 In-degree distribution exponent in web pages of different categories (after Pennock *et al.*, 2002)

	Computer Science	Universities	Companies	Newspapers
γ_{in}	2.66	2.63	2.05	2.05

clusters belonging to different sectors show different connectivity characteristics. For example, Table 14.2 shows the degree distributions exponents of the incoming links to the web pages published online by several groups of organisations.

The study of the topological structure of the web can have quite important 'practical' consequences. A good number of papers published in recent years deal, from both a theoretical and an empirical point of view, with the issues connected to searching and identifying specific elements of the web (Adamic & Adar, 2003; Efe *et al.*, 2000; Flake *et al.*, 2002; Gibson *et al.*, 1998; Lawrence, 2000; Skopal *et al.*, 2003). It is reasonable to predict that the next generation search engines and recommendation systems will extensively use these results to provide their users effective dynamic search tools rather than relying on queries performed on *static* databases (Adomavicius & Tuzhilin, 2005; Almeida & Almeida, 2004; Baggio, 2006; Lawrence, 2000).

The Web Space of a Tourism Destination

Chapter 13 discussed the results of the analysis of the networks of two tourism destinations. In that example we considered the links among websites as connections among the organisations owning them. Let us now focus on one of those, namely the island of Elba, and consider the technological side of the website network. (Baggio *et al.*, 2007a, 2007b). As for many other destinations, the Web has become, in recent years, an important means of promotion and commercialisation for the whole community of local tourism operators. For example, an enquiry conducted on the accommodation entrepreneurs in Elba showed a good acceptance of the internet – 88.8% of the responders had their own homepage and 68.5% had a site within a local tourism portal (Pechlaner *et al.*, 2003). The main motivations for a presence on the network were identified as the possibility of achieving visibility in a wider market and of acquiring new customers, the speed of communication with their customers, but also the passive need to be present on the web.

The websites analysed (468) are the nodes of the network. Hyperlinks among them are considered to be directed links between nodes. The

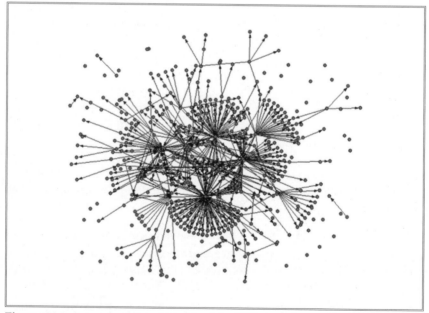

Figure 14.4 A graph of the network of the Elban tourism websites

graphical representation of the Elba tourism websites' network obtained is given in Chapter 13 and is repeated here to ease the discussion (Figure 14.4).

In Chapter 13 we also provided the quantities of the main network metrics. We have seen how the network properties indicate a sensible sparseness (link density d = 0.002 and almost 21% of the websites are isolated nodes). It is, however, a rather 'compact' network (diameter D = 11 and average distance L = 4.5). Moreover, the modularity characteristics are quite limited (global clustering coefficient C = 0.003) and the local (E_{loc}) and global (E_{glob}) efficiencies (Latora & Marchiori, 2001) are 0.0145 and 0.16981 respectively. These values are noticeably lower than those found for similar systems.

Compared with the overall characteristics of the Web (see the values shown in Table 14.1), it is possible to notice that some are in line with those generally exhibited (diameter and average path length), while our example has a much lower degree of modularity (clustering). The low local efficiency value also confirms the poor clustering of the network. The cumulative in-degree (k_{in}) and out-degree (k_{out}) distributions (the distributions of the number of links incoming to a node or outgoing from it) are shown in Figure 14.5. Both display an almost perfect power-law decay $P(k) \sim k^{-\gamma}$. The exponents calculated are: out-degree: $\gamma_{out} = 1.89$ and in-degree: $\gamma_{in} = 2.96$. The in-degree exponent is higher and the out-degree

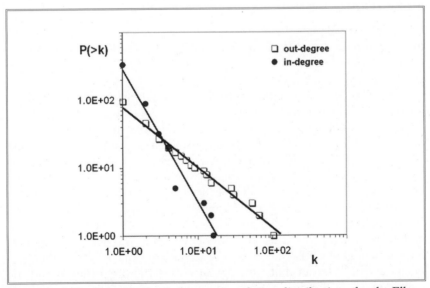

Figure 14.5 In-degree and out-degree cumulative distributions for the Elba tourism websites' network

Table 14.3 Relative size of the components for the Elba network and the web according to the bow-tie model

	Elba network (%)	*Web (%)*
SCC	3.4	28.0
IN	2.1	21.0
OUT	52.4	21.0
TENDRILS	15.6	21.0
TUBE	1.3	9.0
DCC	25.2	

exponent is lower than those typically measured for the Web (see Table 14.1), indicating a greater concentration of the network.

The Elban network, besides the general low connectivity among its websites, still shows evidence of a *bow-tie* structure. Table 14.3 displays the estimated proportions for the components along with the values accepted for the whole Web (Broder *et al.*, 2000). Note also the very low size of a central strongly connected component.

Implications

In summary, the results show a general agreement with those found in the literature on the WWW and the configuration of websites. This reinforces the idea of a substantial self-similarity in the structure of the Web space. However, some of the values for the Elban network show quite different characteristics to that of the Web as a whole: essentially, a much lower connectivity and a higher sparseness of the network.

The first conclusion we may draw from this investigation is that such low connectivity and modularity (i.e.: low and sparse number of connected communities or clusters) is of low efficiency and a waste of resources both from a technical and an organisational point of view. While the benefits of cooperating and collaborating through networking resources and functionalities have been emphasised several times (Barua *et al.*, 2000; Hackathorn, 2003; McLaren *et al.*, 2002; Walker, 2002), in the destination examined here this is not evident in connections between websites. Elba appears to be missing out on the advantages of cooperation mechanisms on the Internet that could greatly ease the organisation and the management of the destination and its efficiency in facing a globalised and highly competitive market.

The tourism organisations in Elba, as in most European countries, are mostly small and medium-sized enterprises. Strong collaboration and cooperation practices in the field of information technologies could be greatly beneficial for these companies, helping them to overcome many of the problems they face in addressing the effective and efficient use of ICTs (typically: skills, infrastructures and costs). Forming a critical mass of users, sharing costs and resources, the development and the deployment of sophisticated systems for the management or the operations (yield or revenue management, administrative and financial management, relationships with intermediaries etc.) would not be demanded to single entities, which might not have sufficient means to tackle such projects. On a more general destination management level, collaboration among the stakeholders may mean, for example, more efficient collection of the data needed to plan and coordinate the destination activities, or harmonisation of reception capabilities (via a centralised booking system), improving the satisfaction of the tourists.

Today, in fact, most advanced international destinations are able to sell integrated products, and individual tourists can satisfy their informational or operative needs by navigating through well-interconnected groups of websites. Elba, instead, shows a very a scarce connectivity and a high rate of disconnected elements. This means, for example, that a guest has to identify a single supplier's website address in order to compose a holiday package. More importantly, this is also a clear sign that the entrepreneurs are not cooperating to offer integrated products and services.

The second important consideration pertains to a strategic development perspective. As discussed above, future search engines and recommendation systems will be based, with high probability, on dynamic agents whose task will consist in identifying connected communities on the web. The websites of destinations not forming an identifiable 'community' through high network interlinkages will be hard to reach by a casual user, with unfavourable consequences for the effectiveness of the marketing and communication activities and their economic outcomes.

A presence on the Web is today an important factor of competitiveness. At a destination level, the necessity to implement and develop collaboration or cooperation mechanisms has been stated several times (Bramwell & Lane, 2000; Jamal & Getz, 1995). Besides the more specific considerations noted above on the advantages of technical cooperation, the capability to identify oneself as a cohesive web community may be, in the future, an important discriminator even for the recognition of existence of a tourism destination. As Bernat *et al.* note:

Web sites that are commercial successes for the individual organisations behind them, but which contribute nothing vital to the physical or virtual communities in which they are embedded, must be considered rhetorical failures. Alternatively, cooperative collections of stakeholders' sites that are rhetorically sensitive to their effects on the whole community system will undoubtedly facilitate graceful transitions to the digital age of communication. (Bernat *et al.*, 2003: 35)

Simulation of Tourism Networks

In the previous paragraphs the results of the Elba network analysis have allowed us to describe a model for the destination examined. A model is one of the fundamental instruments of modern science. A great part of the efforts of a scientist goes into developing, assessing and improving models. Fundamentally this is because:

> No substantial part of the universe is so simple that it can be grasped and controlled without abstraction. Abstraction consists in replacing the part of the universe under consideration by a model of similar but simpler structure. Models formal or intellectual on the one hand, or material on the other, are thus a central necessity of scientific procedure. (Rosenblueth & Wiener, 1945: 316)

Different types of models exist. They can be sets of equations describing the behaviour of an element or system (Newton's laws, for example), mechanical devices aiming at easing the intuition of basic principles (such as the 18th century orreries, representing the motion of solar systems) or even mental models, used to help the foundation of formal theories (such as the Einstein's idea of a beam of light which eventually led him to the theory of relativity) or to refute existing ones (see, for example, Galileo's discussions in refutation of Aristotelian ideas on motion).

The power of a model resides in its capacity not only to describe a system, but also to allow 'playing' with it, exploring different possibilities or different configurations; in other words, to perform simulations and to derive useful predictions on some characteristic of the system.

Obviously, mathematical (analytical) models have been always considered the best choice to achieve these results. The idea of using some kind of mathematical relationship to represent a natural system or phenomenon is as ancient as the history of science. Equally ancient, however, is the recognition that ideal cases, in which a well-defined set of equations can describe completely even simple arrangements, are almost non-existent. The *Mathematike Syntaxis*, better known as the *Almagest*, by Ptolemy (1st

Figure 14.6 An 18th-century orrery: a mechanical model of the solar system

century), is a wonderful geometrical construction explaining the motion of heavenly bodies. In order to give practical solutions, however, Ptolemy was forced to provide a series of numerical tables, so that his book is more an astronomical almanac than a coherent theoretical discussion of the celestial motions. Some centuries later, the work of Poincaré on the three body problem (1883, 1884), showed beyond doubt that even the simplest Newtonian systems involving more than two bodies may exhibit very complicated dynamics and almost unpredictable results may arise from small variations of the initial conditions.

A vast part of the work of the mathematicians through the ages has been devoted to the development of approximate methods to find a practical solution to these problems. They have strived to find practical ways to solve otherwise unsolvable problems and to simulate the behaviour of systems and phenomena that would have been otherwise impossible to investigate. From a practical point of view, however, only the advent of electronic computers has provided the opportunity to actually perform such simulations. Numerical methods invented by precedent mathematicians were immediately transformed into computer programs and used in a wide number of disciplines. Probably the best-known of all is the *Monte Carlo* method, proposed by Metropolis at Los Alamos during the Manhattan Project. His algorithm (Metropolis *et al.*, 1953) is considered today the one 'with the greatest influence on the development and practice of science and engineering in the 20th century' (Dongarra & Sullivan, 2000).

In addition to providing the opportunity to describe and explain phenomena, numerical simulations allow experiments to be performed in fields where these would not otherwise be feasible for theoretical or practical reasons. We have seen in the introduction to this book that a network is a system which may comprise a very large number of elements. We have seen also that the topological characteristics of a network have a direct relationship with many dynamic processes. It is therefore interesting to experiment with different configurations to measure these effects or to understand better how some particular composition may influence the behaviour of the whole system.

On the other hand, we may want to find the optimal (the most efficient, the least expensive, the fastest) configuration able to sustain or to react to a certain process. If the network represents a social system, for example, it would be impossible to test the hundreds or thousands of different possible configurations in the field, measuring, for all of them, the quantities we are interested in (Edmonds & Chattoe, 2005).

A numerical simulation is, therefore, the only possible solution to such problems, and a numerical simulation means, today, a computerised simulation. In what follows, by simulation we mean the design, the implementation and the execution of computerised algorithms (computer programs) that (attempt to) reproduce an abstract model of a particular system.

Computer Simulations in the Social Sciences

Social scientists have been using simulation techniques for a long time (Inbar & Stoll, 1972). The wide availability of computing power and of efficient programming languages, coupled with much simpler access to data has, in recent decades, greatly increased the attention to these methods and their practical uses (Conte *et al.*, 1997; Gilbert, 1999; Suleiman *et al.*, 2000). A growing body of literature and some dedicated publications (see the *Journal of Artificial Societies and Social Simulation*, for example) have been accumulating in recent years. They are providing examples, investigating techniques and strengthening the theoretical bases on which these experiments are built. According to the reviews by Halpin (1999) and Axelrod (2006), the most important uses of simulations are in the areas of:

- *Prediction*: where complicated inputs and non-analytic processes can be numerically combined in order to infer some characteristic or behaviour in future times (Bankes, 2002).
- *Execution*: a typical application is the usage of artificial intelligence techniques to perform certain operations that would be too difficult, long or complex for a human operator, from the analysis of texts and discourses to the recognition of patterns, images or speech (Bezdek *et al.*, 2005; Conte *et al.*, 2004).
- *Training and education*: a fast-developing area in which a simulated environment (very often taking the form of a computer game) is used to provide, in an effective and accelerated way, different possible scenarios to educate people in unusual or potentially dangerous conditions. Probably the best examples in this field, although a little off-topic, are flight simulators (Maier & Größler, 2000; Prensky, 2001).
- *Verification and discovery*: using simulations as scientific method-ological tools, they can provide insights into relationships, behav-iours, properties of complex dynamic social systems. This is the case of some recent theories on the formation of opinions (Stauffer, 2003, 2004; Sznajd-Weron & Sznajd, 2000), the diffusion of ideas (Da Costa & Terhesiu, 2005) or the dynamics of culture propagation (Axelrod, 1997).

These techniques have already proved able to provide successful answers to both theoretical and practical problems such as the explanation of the distribution of political votes (Bernardes *et al.*, 2002), or the issues involved in forecasting the impacts of policy options in the management of tourism activity and development (Walker *et al.*, 1998).

More complex systems are being developed. One example is the Hydra prototype (Birkin *et al.*, 2005), a spatial decision support system designed to represent the entire UK population. Its set of modelling tools allows us to address specific research or policy questions, from the distribution of facilities for care around an urban area to the problems connected with emergency vaccinations in response to epidemics such as SARS or even smallpox (Barrett *et al.*, 2005).

The literature on simulation and social sciences also expresses some concerns regarding a too wide acceptance of these methods or their gener-alisation capabilities. Edmonds (2004), for example, while recognising a theoretical validity to simulations, warns that their main characteristics are experimental in nature. He notes that in any computer simulation, the formal representation of phenomena and agents is at risk of distortion and that it would be impossible to have a full understanding of the outcomes of a simulation in all possible circumstances. Due to their complexity and their characteristics, computer-based models are also subject to some limitations and pitfalls and some results may be a deterministic consequence of their mathematical structure rather than uncovering the emergence of a complex system's properties (Bertels & Boman, 2001).

On a more general basis, it has also been argued that the supposed superiority of non-linear models might be apparent (Richardson, 2003: 1): 'The presumption is that because these models are more complex than their linear predecessors they must be more suited to the modelling of systems that appear, superficially at least, to be (compositionally and dynamically) complex'.

Moreover, in many cases the results of simulations are then used for deterministic linear predictions or decision-making activities, thus enhancing some of their possible intrinsic weaknesses (excessive simplification of underlying models, for example). In this respect, the attitude towards a model is crucial (Richardson, 2003: 8): 'The linear "culture" takes a representationalist view of models in which aspects of reality really are considered to be captured by the model itself – the model becomes an accurate map of reality á la realism. Even if the model itself is nonlinear its efficacy tends to be overstated'.

Verification and Validation of Computer Simulations

A good simulation can be used to analyse the behaviour of a system under different conditions and can be used to make decisions as if it would have been possible to experiment with the system itself (Law & Kelton, 2000). Obviously this is true only if the simulation used can be 'validated' in some way. The problem has been discussed several times by academics and practitioners from the very beginning of the 'computer age' (Conway, 1963; Conway *et al.*, 1959; Naylor & Finger, 1967). Apart from the debate on the epistemological meaning of simulations, the literature on the subject is quite extensive (Henrickson & McKelvey, 2002; Klein & Herskovitz, 2005; Kleindorfer *et al.*, 1998; Schmid, 2005).

From a general perspective, there are no standard theories or toolboxes for validating and verifying a computer simulation model (Kleijnen, 1995). Therefore, as happens in similar cases, only a rigorous methodology can provide the desired results. The literature offers many possible alternatives and variations, but, essentially, focuses on a path which may be summarised as follows (Kleijnen, 1995; Küppers & Lenhard, 2005; Law & Kelton, 2000; Naylor & Finger, 1967; Stanislaw, 1986).

First of all, let us define validation as 'the process of determining the sufficient accuracy to which a model or simulation is a representation of the real world system from the perspective of the specific purpose of the model or simulation' (Schmid, 2005: 4.1). This implies that no absolute value can be given to such a process. Its value will be dependent on the specific situation or the specific purpose.

The validation process starts with the design of a conceptual model. This must be checked in order to assess its adequacy in representing the system considered. Its assumptions should be tested empirically or by referring to other models known in the academic literature or among expert practitioners. This is then transformed into a computer program. Here, standard practices and techniques exist to verify all aspects of correctness, usability and usefulness of the software developed (Adrion *et al.*, 1982; Balci, 2003; Bérard *et al.*, 2001). Finally, the results of the simulation are tested to determine their correctness and suitability to the situation under investigation. This can be done by comparing these results with available analytical results or with available real responses of the system. Appropriate statistical procedures and tests allow performance of this assessment.

The whole validation process must also guarantee the best possible replicability of the simulation itself and of the results. A good *credibility* is also needed for a simulation to be accepted by a decision-maker. It must be noted here that credibility and validity do not necessarily imply each other. While a validated simulation derives credibility from this process, a simulation can be thought credible based only on the comprehension, the involvement or the agreement with its hypotheses, or the reputation of the developers. These characteristics, however, although important, are not sufficient to validate a simulation.

Networks seem a natural candidate for computer simulations. They can be modelled with sound mathematical techniques (graph theory) and they are a sufficiently general object with which we represent a wide variety of natural and artificial systems of different complexity. Moreover, as has been seen in earlier chapters, many dynamic processes can be defined and studied, all of them of interest to scientists and practitioners. The great majority of the theoretical results in this field come from computer simulations. In many cases the use of a simulation approach to model and analyse a network or to predict its behaviour has proved very important, not only to gain a better understanding of such systems, but also (computer or transportation networks, for example) to design, deploy, and manage them.

Software Tools

The great increase in network studies has also induced the availability of a wide range of computer programs to aid in analysing and visualising networks. The web page 'Computer Programs for Social Network Analysis' published by the INSNA (International Network for Social Network Analysis at: http://www.insna.org/INSNA/soft_inf.html contains more than 50 references to software programs which cover a wide variety of functions. Among these, probably the most used and known are Pajek (http://vlado.fmf.uni-lj.si/pub/networks/pajek/) and UCINET (http://www.analytictech.com/).

Most of these programs, however, while providing extensive sets of tools and functions to estimate the many statistical parameters of a network, have only limited simulation capabilities. They typically limit themselves to the capability of generating different kinds of networks, with the possibility of specifying in advance some feature such as size, degree distribution type or link density.

Different simulations must be programmed individually. The first and most obvious possibility is the usage of a standard programming language (C++, Java, Delphi, Fortran), perhaps with the aid of specific libraries (examples are the Boost Graph Library for C++ at http://www.boost.org/, or JUNG – the Java Universal Network/Graph Framework at http://jung.sourceforge.net/).

A further possibility is the usage of high level languages (which are sometimes also called *development environments*) such as Matlab (http://www.mathworks.com/), Maple (http://www.maplesoft.com/), Mathematica (http://www.wolfram.com/), and Gauss (http://www.aptech.com/). These are characterised by built-in capabilities for many basic functions (graphics, matrix manipulation, mathematical functions) and much higher flexibility in handling data structures. The advantage is that the user can focus more easily on the solution of a problem rather than coping with the specificities of a programming language. The drawback is that there may be some limitations in speed and capacity when handling large quantities of data. They are also generally commercial (and quite expensive) packages. Some of them, however, have freeware clones which possess very similar language structure and functionalities. Examples are: Octave (http://www.octave.org/) a Matlab clone, or Ox (http://www.doornik.com/) which is quite similar to Gauss.

One more environment dedicated to simulations is a class of toolkits developed to implement agent-based models (ABM). The basic idea of such simulations is that a system is composed of a number of entities (agents) which behave according to some simple rule (Flake, 1998; Wolfram, 2002). The interactions of the agents can give rise to some global system property which can then be examined. Variations in the basic rules or in the typology of the agents generate different final configurations for the system. These models have been shown to be quite effective and efficient in simulating different types of social and natural systems and processes, and may prove very valuable as an aid in decision making (Tesfatsion & Judd, 2006; Toroczkai & Eubank, 2005).

Again, being ABM computer programs, a traditional language can be used. However, in recent years, a number of dedicated programs have been developed to help with ABM simulations. Software packages such as Swarm (http://www.swarm.org/), RePast (http://repast.sourceforge.net/), NetLogo (http://ccl.northwestern.edu/netlogo/) and many others provide libraries with functionalities at different levels of complexity. Some of them are relatively easy to use and fast to learn; some require much greater effort and specific programming skills. Typically, software that is easier to learn and use is less powerful or more rigid, leaving less flexibility for the user.

It is interesting to note that, with the exceptions seen above, the vast majority of these software packages are provided free of charge on the Internet. And, even in case of commercial products, there is a wide availability of pre-programmed (and sometimes very sophisticated) routines which can greatly help a rapid development of simulation models.

A computer simulation, in general, is implemented by using a series of tools such as those described above. The following discussion gives some practical examples.

An Example: The Network of a Tourism Destination

Let us consider the diffusion of a rumour (or information, fad, policy message) in a real network and let us see its influence on the network topology. The network considered is the one formed by the stakeholders of a tourism destination: the island of Elba, Italy. In Chapter 13 we described the main topological characteristics of this network. We have seen that it exhibits a clear scale-free structure (a few hubs with many connections and many nodes with very few links), a low density of connections and a very limited clustering (Baggio *et al.*, 2007a; Baggio *et al.*, 2007b). The degree distribution of this network, a typical feature of a complex network, is a power-law distribution with an exponent (considering the network undirected) $\gamma = 2.18$, which is lower than the one usually exhibited by such systems, i.e. the distribution of the connections is flatter and more sparse.

The conceptual model used for the analysis of the rumour diffusion is a simplified version of the one described, for example, by López-Pintado (2004) or Pastor-Satorras and Vespignani (2001). At a time t_0 one of the elements of the network starts spreading a rumour. It does so by 'telling' it to all the nodes connected to it. We are interested in studying this phenomenon. The results are compared with those obtained by considering two synthetic networks of similar size and link density: one with a random distribution of the links (random), the second one (scale-free) with a pure scale-free topology (exponent of the degree distribution: $\gamma \sim 3$). In our simulation we take into account only the giant cluster of the networks, the main fully connected component.

The diffusion model has been implemented with Netlogo (Figure 14.7) and is a derivation of some of the distribution library models (Rumor Mill as modified by F. Sondahl http://www.cs.northwestern.edu/~fjs750/netlogo/). The model allows the variation of a number of parameters: the initial rumour can be spread by a single element or by a certain number of them; at each step a node can tell the rumour to all of its neighbours or only

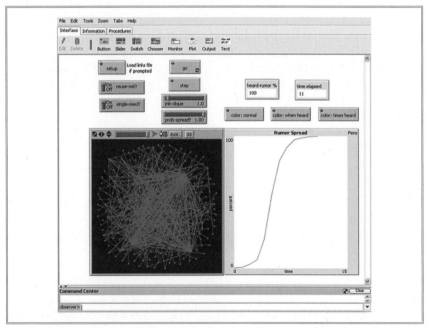

Figure 14.7 Diffusion simulation model: the Elba network at the end of a simulation run

to a certain percentage of them. At the end of a simulation run, it is also possible to visualise how often the nodes have been told the rumour or how recently they have been aware of it.

After a series of runs, the diffusion curves shown in Figure 14.8 are obtained (the curves are averages after ten realisations). As it can be seen, the scale-free networks are much more 'efficient' in the diffusion process. The total times needed to inform the whole network are 9 for the scale-free network, 11 for Elba and 14 for the random network.

As expected, both scale-free networks respond in a faster way to such a process. We may also notice a clear difference between our empirical Elba network and a pure scale-free system. Elba's flatter degree distribution translates into a lower efficiency of the network compared with the other. By changing the parameters of the model we may also continue this experiment and find different possible configurations. For example, we find that the diffusion in Elba can reach the speed of the scale-free network if it is initiated by more than a single element (5% of the number of nodes will suffice to achieve this goal).

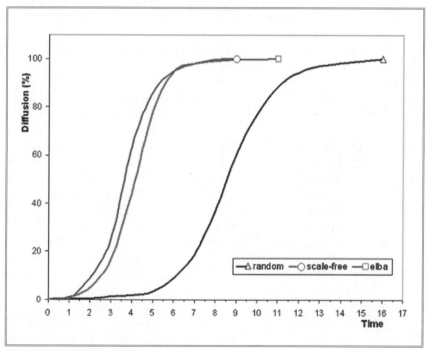

Figure 14.8 Diffusion curves

From these results it is possible to find a way to achieve a better diffusion: we may change the topology of the network by adding links in a way to obtain a steeper degree distribution or, if this is not feasible, by recognising that to spread a rumour we should start it not from a single source but from a certain number of *rumour-spreaders*.

Once the diffusion process has been studied, we may want to know something more. For example we may want to investigate the robustness of the network. In other words, we want to know what may happen to the connectivity of the network if a certain fraction of the connections is removed. To do that, it is possible to simulate a random removal of the connections and compute some parameter which can give us meaningful measure. The representative quantity is the efficiency (both local and global) of the network, interpreting it as a measure of the capability of the system to exchange information over the connections. The local efficiency refers to the single nodes, the global efficiency measures the property, at a large-scale level, for the whole network. The simulation applies the model and the procedure described by Albert *et al.* (2000) and Latora and Marchiori (2001). We start with the whole network and compute the local E_{loc} and the global E_{glob} efficiencies. At each step a random number of links (5%) is removed and the parameters calculated again. As in the previous case, the results are compared with a similar simulation performed on a random network. The simulation, in this case, has been programmed with the Matlab development environment.

The results are shown in Figures 14.9 and 14.10. In both, f is the fraction of the original networks and E_{ind} is the efficiency value calculated as a ratio over the value for the entire network; the efficiency values are the averages over ten realisations. As would have been expected (scale-free vs. random networks) the Elba system shows a higher robustness in the case of random removals of connections. In particular, there is a sudden transition in E_{loc} at $f \sim 75\%$ for the random network. At this value the network is completely disrupted.

It must be stated however, that the difference in the behaviours of the two networks is not as large as it would have been expected (Crucitti *et al.*, 2003). This, again, can be explained by considering the characteristics of the Elba network and particularly the fact that a *flattish* degree distribution makes it closer to a random network. Our network is more robust than a random network, but its robustness is quite fragile. In this case, theory and models predict that the only way to achieve a better capability of the Elba network to resist to random shocks that might reduce its information transfer efficiency is by a thorough reconsideration of the connectivity characteristics.

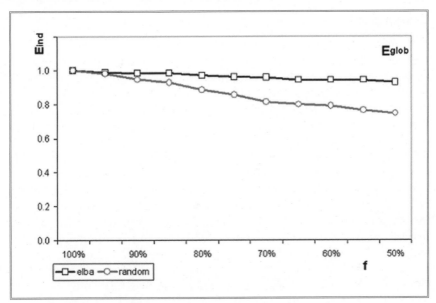

Figure 14.9 Simulation results: global efficiency

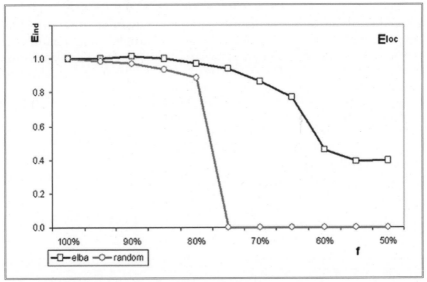

Figure 14.10 Simulation results: local efficiency

Conclusion

In this chapter we have discussed the importance of ICTs and technological networks for the tourism sector and demonstrated the potential of network simulation. We have outlined the most recent and important results dealing with the study of the web as seen from the network science viewpoint in the framework of complex systems science. An example of the usage of these methods in the analysis of a real case and the results of the analysis have been presented. This line of investigation is at a very initial stage of development and very few cases have been studied. Still, some interesting conclusions have been deduced, both from a theoretical and an applied point of view. More work is obviously in order before being able to generalise what is discussed here. More cases need to be investigated before being able to draw a general picture of the salient features of tourism systems' website networks and before being able to state if and how they are different from the global Web. Future research also needs to deepen the analysis by considering not only a static snapshots of these systems, such as the one presented here, but also their dynamic evolution, and, possibly, to connect it to the current understanding of the history and the development of a tourism destination.

We have also shown that computer simulations are important tools in the study of a complex network. Provided a good methodology is used to assess the validity of the model used and of its realisation, this can be the only possible way, in many cases, to gain useful insights from the analysis of a network, of its behaviour in particular conditions and of the processes that can occur across it. The examples discussed, although at an elementary level, provide evidence of this possibility.

Chapter 15
Conclusions

Introduction

This book has provided an overview of the existing literature on network analysis and tourism and mapped a body of knowledge that extends across concepts such as social capital, governance, leadership, policy and the dynamics of change and power. This disparate literature is united by the concept of tourism networks of stakeholders within a destination. However, this book is not the final work on tourism network analysis nor has it sought to be. Instead it provides an indication of the usefulness and importance of network thinking in tourism research and practice, a discussion of the various approaches that have been used to study tourism networks, an introduction to a more numerate and quantitative style of research and how, by embracing this style, researchers can open an enormous range of opportunities to study tourism phenomena using theory that has proven useful in other disciplinary areas. Nonetheless, network analysis is not the total answer to the study of tourism phenomena; indeed as this book has shown, qualitative or quantitative approaches to the study of networks each provide only a partial view of reality. However, the use of quantitative network thinking has proven apposite in a variety of research areas and the authors consider it a useful addition to the tourism researchers' toolkit. It is therefore pertinent to ask why more researchers in tourism do not use such a quantitative approach.

Firstly, network analysis focuses on relationships and the effect of the structure of relationships on individual behaviour. As we have argued in this book, thinking of relationships rather than individual actors represents a paradigmatic shift in the unit of analysis for research and is at odds with much of classical management, economics and sociological thought. It therefore may challenge researchers whose work is embedded in these literatures, and have an existing pattern of academic relationships and thinking that may determine their research activity. Perhaps then we should look to a younger generation of tourism researchers to adopt techniques of network analysis.

Secondly, as this book has demonstrated, quantitative network analysis requires mathematical skills and the mastery of a specialist vocabulary, including words such as ego, alter, node, arc, edge, and tie; network

217

properties such as centrality and density; and analysis techniques such as blockmodelling. These concepts, techniques and properties do require an effort to understand them and it may be more than some established tourism academics may wish to undertake. Nonetheless, we feel that this effort is worthwhile, given the insights that network analysis can provide for tourism researchers.

Thirdly, it may also be that the structural 'turn' and the idea that people's behaviour is determined in part by relationships and organisational structures and is therefore beyond the complete control of an individual, is philosophically anathema to some tourism researchers. Perhaps these researchers should pass by and avert their eyes from the ideas in this book.

Finally, it may be that some tourism researchers believe that network analysis is atheoretical and simply a diverse and incoherent series of tools. Perhaps it is these people which this book has sought to influence most. It is clear to the authors that the chapters of this book provide clear evidence that many of the key concepts and theories of tourism are informed by network analysis.

Directions for Future Research

A number of authors have provided suggestions for further research relating to network analysis (Parkhe *et al.*, 2006). The focus of this section is instead on how network analysis may be further applied in the area of tourism research. The chapters of this book have illustrated how many of the key concepts for tourism research, such as knowledge management, governance, destination performance and social capital, appear related through the concepts of stakeholders and networks. Thus network analysis, especially quantitative network analysis, has the potential to inform all of these areas. A range of future research directions for tourism, filtered through the specific interests of the authors, are examined below.

In understanding how the ideas of this book may be used by other researchers, it is important to think about access to data and information. Another reason that may explain the lack of quantitative network analysis in tourism studies is that it may be difficult to access the requisite information. Many types of quantitative network analysis require complete sampling of people in a particular domain and this may be a challenging, and expensive, task. There is a stream of research regarding network sampling that may address this issue (see for example Granovetter, 1976) but access to respondents can be difficult. In part this is driven by the need for the research outcomes to be seen as important and useful to the tourism industry. Clearly tourism is an area where organisational efficiency and effectiveness is critical for areas such as destination management and marketing. However, it is also an area where personalities and politics play an important part in decisions and where network analysis may uncover inefficiencies in the organisation of the actors who do not want this information made available. This issue could be addressed by increasing the acceptance of network analysis as a tool in enhancing the efficiency of tourism destinations. One route to achieve this would be to encourage comparative studies of tourism destination organisation using network analysis as an area of emphasis for future research.

A further example where quantitative research may be useful is in the area of policy network analysis. Most policy network analysis in tourism is qualitative in nature and does not make use of quantitative network analysis techniques. One prominent exception is Pforr (2002) who has analysed tourism policy processes using network analysis. It would appear that the policy network concept has significant further potential for using quantitative techniques for measuring and analysing the relationships among destination stakeholders.

Concepts Related to Organisational Structures and Systems

One promising area of research for tourism researchers is to adapt or influence wider inter-organisational network thinking through examining the applicability of concepts developed in other areas to the study of destination management. Prior authors such as Borgatti and Foster (2003) have examined the consequences of network influences in organisational research indicating that social capital explains performance variations amongst social groups. Clearly, the effectiveness of a tourism destination requires operators to work together. One of the maxims of tourism business is 'Sell the Destination First' and refers to the imperative of developing a common message and approach amongst a disparate set of tourism operators within a destination. In the past, tourism researchers have also examined this idea using the concept of social capital (Jones, 2005). An alternative conceptualisation involves the notion of cohesion. Cohesion researchers state that individuals are influenced by their direct ties because of the frequency, intensity, and proximity of interaction (Burt, 1987). Restricted information exposure and conformity pressures within cohesive groups also influence cohesive actors (Levine & Moreland, 1990). It may well be that development of a cohesive group of destination operators may be enhanced through examination of these factors using network analysis and in turn clarify the applicability of the social capital concept to groups of businesses.

The concept of tourism destinations as systems is widespread in the tourism literature and is based on the idea of interaction between the different parts of the tourism system (Leiper, 1989; Becken, 2004; Laws, 2003; Carlson, 1999). Some authors have argued that this interaction is chaotic or may lead to chaotic situations (Faulkner, 1997; Russell, 1999; McKercher, 1999; Faulkner, 2001). Network theory provides an alternative view of a tourism destination as a series of relationships between businesses which link various parts of the tourism system. In this view then, chaos and change in a tourism destination system would be marked by changes in the relationships among the operators involved in the destination. Here network thinking may provide an alternative conceptualisation of the dynamics of the development of destinations.

The idea of destination networks as systems prompts another area where there is potential for tourism research. This lies in the area of exchange of information amongst the network of stakeholders within a destination. Such information exchanges are vital for the efficient management of a destination and help to develop a distinct culture and trust among members as well as to stimulate effective communication between organisations such as the regional tourism office and its members. Thus network analysis may help to determine where there are weaknesses in information flow within a destination (Scott, 2006). Again, the development of standardised methods of network analysis would provide the opportunity to compare and contrast different destinations and their communication effectiveness.

However, network analysis is useful not only within destinations but also to examine linkages between tourism and other sectors of the economy. For example, prior research has shown that knowledge management is an important element in the management of tourism destinations (Cooper, 2006). Network analysis based upon exchange of information offers the opportunity to diagnose where linkages between actors may be enhanced (for example between academia and tourism operators).

Complexity and Tourism

Finally, this book has introduced, through the work of one of the authors, Rodolfo Baggio, some approaches to the study of social networks using techniques derived from such areas as scale-free networks studied in the physical sciences. While the intriguing possibility of universal principles for the study of social systems has drawn the attention of scholars in the past, the application of tools and techniques taken from the physical sciences to the study of the Internet and social systems has seemingly uncovered some interesting commonalities. For example, Baggio (2007b) has contributed to the strengthening of the idea that *complexity* is a distinctive characteristic of a tourism system.

Within this broad framework, network analysis methods have found similarities between the characteristics of tourism Internet site linkages and scale-free networks and have highlighted the possibility of using these new techniques to improve our knowledge of the structure and the behaviour of a tourism system (Baggio, 2007a). Similarly, this approach has revealed that it is possible to compare the properties of Internet networks for tourism destinations (Baggio *et al.*, 2007). The results, along with the possibilities provided by the application of computerised numerical simulation methods, have shown their effectiveness and usefulness for both the researcher interested in the theoretical study of a tourism destination, and the practitioner in need of more sophisticated instruments to cope with the challenges of fostering social and economic developments in the complex and dynamic environments that are destinations.

Further research in this area will first need to confirm the results obtained so far by increasing the number of examples studied. The methodology proposed requires some further refinement both from a practical and a theoretical point of view. Different paths are open, but the most promising looks to be the one related to the investigation of the evolution of a tourism destination and the relationships that lie within the structure of the web of linkages among the tourism operators. This will include issues such as the extent and the forms of cooperation and collaboration found in a destination. The literature has already identified these as important success factors, and they can be better assessed by combining traditional qualitative investigation techniques with the quantitative approach presented here. Once assessed, the network structure can then be the starting point of a series of simulation exercises which can provide useful insights able to shape future scenarios and strategies and inform possible management actions. Models and tools for these activities will probably be one of the most important development areas in this field for the next years.

In other words we should not see destination networks as static structures, but rather the techniques and approaches described above offer the possibility of informing the dynamics of network change. This is an area of particular interest for physicists and indeed almost all of the hundreds of articles on networks contributed by physicists in the last few years have focused on the evolution of networks (Newman, 2002). There is a small but developing area of research that may be termed organisational network change (Gulati & Gargiulo, 1999; Madhavan *et al.*, 1998; Shah, 2000) and a significant amount of recent work on inter-organisational networks is about explaining how and why organisations form ties and select partners. Similarly, the large literature on the effects of proximity and homophily is about what causes networks to develop (McPherson *et al.*, 2001), and recently there have been attempts to model individual behaviour based on networked embeddedness (Macy & Skvoretz, 1998). For tourism, these topics related to network dynamics constitute virgin territory. There is virtually no research on the evolution of destination networks, despite the considerable literature on the evolution of destinations. There is a real need here for research on both 'back-casting' and 'fore-casting' destination networks.

In conclusion, the concept of networks and network analysis fits naturally with the study of tourism and tourism destinations. Tourism destinations naturally lend themselves to be conceptualised as networks and the approach offers researchers the opportunity to study destinations in a way that can potentially contribute significantly to the wider literature and our understanding of how tourism functions. Perhaps finally, in this way, tourism can become a leading area for study rather than a follower of research trends in other areas.

References

Aaker, D.A. (1991) *Managing Brand Equity: Capitalizing on the Value of a Brand Name.* New York: Free Press.

Aaker, D.A. (2004) *Brand Portfolio Strategy: Creating Relevance, Differentiation, Energy, Leverage, and Clarity.* New York: Free Press.

Aaker, J.L. (1997) Dimensions of brand personality. *JMR, Journal of Marketing Research* 34 (3), 347–56.

Acs, Z.J. and Audretsch, D.B. (1988) Innovation in large and small firms: An empirical analysis. *American Economic Review* 78 (4), 678–90.

Acs, Z.J., and Audretsch, D.B. (1993) Innovation and firm fize: The new learning. *International Journal of Technology Management* 8 (2), 23–35.

Acs, Z.J. Audretsch, D.B. and Feldman, M.P. (1994) R and D spillovers and recipient firm size. *Review of Economics and Statistics* 76 (2), 336–40.

Adamic, L.A., and Adar, E. (2003) Friends and neighbours on the web. *Social Networks* 25 (3), 211–30.

Adams, K. (2002) At the table with Arendt: Toward a self-interested practice of coalition discourse. *Hypatia* 17, (1–33)

Adler, N.J. and Graham, J.L. (1989) Cross-cultural interaction: The international comparison fallacy? *Journal of International Business Studies* 20, 515–37.

Adler, P.S. and Kwon, S.W. (2002) Social capital: prospects for a new concept. *Academy of Management Review* 27 (1), 17–40.

Adomavicius, G. and Tuzhilin, A. (2005) Toward the next generation of recommender systems: A survey of the state-of-the-art and possible extensions. *IEEE Transactions on Knowledge and Data Engineering* 17 (6), 734–49.

Adrion, W.R., Branstad, M.A. and Cherniavsky, J.C. (1982) Validation, verification, and testing of computer software. *ACM Computing Surveys* 14 (2), 159–92.

Agger, A. and Löfgren, K. (2006) *How Democratic are Networks based on Citizen Involvement?* Roskilde: Centre for Democratic Network Governance.

Agostinho, M.E. and Teixeira de Castro, G. (2003) Co-creating a self-organizing management system: A Brazilian experience. Paper presented at the Complexity, Ethics and Creativity Conference, London School of Economics (17–18th Sept. 2003)

Ahuja, G. (2000) Collaboration networks, structural holes, and innovation: A longitudinal study. *Administrative Science Quarterly* 45 (3), 425–55.

Alba, RD (1982) Taking stock of network analysis: A decades' results. *Research in the Sociology of Organisations* 1, 39–74.

Albert, R. and Barabási, A. (2002) Statistical mechanics of complex networks. *Reviews of Modern Physics* 74(1), 47–97.

Albert, R., Jeong, H. and Barabási, A.-L. (2000) Error and attack tolerance of complex networks. *Nature* 406, 378–82.

Alford, P. (1998) Positioning the destination product – Can regional tourist boards learn from private sector practice? *Journal of Travel and Tourism Marketing* 7 (2), 53–68.

223

Almeida, R.B. and Almeida, V.A.F. (2004) A community-aware search engine. *Proceedings of the 13th International Conference on World Wide Web (WWW2004)*, New York, NY, USA, 413–21.

Amaral, L.A.N. and Ottino, J.M. (2004) Complex networks – Augmenting the framework for the study of complex systems. *The European Physical Journal B* 38, 147–62.

Amaral, L.A.N., Scala, A., Barthélémy, M. and Stanley, H.E. (2000) Classes of small world networks. *Proceedings of the National Academy of the Sciences of the USA* 97, 11149–52.

Amin, A. and Thrift, N. (1994) *Globalisation: Institutions and Regional Development in Europe*. Oxford: Oxford University Press.

Anderson, P.W. (1972) More is different. *Science*, 177 (4047), 393–6.

Andsager, J.L. and Drzewiecka, J.A. (2002) Desirability of differences in destinations. *Annals of Tourism Research* 29 (2), 401–21.

Antonioli Corigliano, M. and Baggio, R. (2006) On the significance of tourism website evaluations. In M. Hitz, M. Sigala and J. Murphy (eds) *Information and Communication Technologies in Tourism 2006 – Proceedings of the International Conference in Lausanne, Switzerland* (pp. 320–31) Wien: Springer.

Arabie, P. and Wind, Y. (1994) Marketing and social networks. In S. Wasserman and J. Galaskiewicz (eds) *Advances in Social Network Analysis* (pp. 254–73). Thousand Oaks: Sage Publications.

Araujo, L. and Bramwell, B. (2000) Stakeholder assessment and collaborative tourism planning: The case of Brazil's Costa Dourada project. In B. Bramwell and B. Lane (eds) *Tourism Collaboration and Partnerships: Politics, Practice and Sustainability* (pp. 272–94). Clevedon: Channel View.

Archdale, A. (1995) New frontiers for tourism technology. *Insights: Tourism Marketing Intelligence Service*, 6 (9), D27–31.

Arendt, H. (1958) *The Human Condition*. Chicago: University of Chicago Press.

Argote, L. and Ingram, P. (2000) Knowledge transfer: A basis for competitive advantage in firms. *Organizational Behavior and Human Decision Processes* 82, 150–69.

Argote, L., Beckman, S.L. and Epple, D. (1990) The persistence and transfer of learning in industrial settings. *Management Science* 36 (2), 140–54.

Armstrong, D. (1983) *The Political Anatomy of the Body*. Cambridge: Cambridge University Press.

Armstrong, H. and Taylor, J. (2000). *Regional Economics and Policy* (3rd edn). Oxford: Blackwell Publishers.

Arrell, R. (1984) *Waitomo Caves: A Century of Tourism*. Waitomo Caves Museum Society, Waitomo Caves.

Arthur, W.B., Durlauf, S. and Lane, D. (eds) (1997) *The Economy as an Evolving Complex System II*. Reading, MA: Addison–Wesley.

Atkinson, M. and Coleman, W. (1992) Policy networks, policy communities and the problems of governance. *Governance: An International Journal of Policy and Administration* 5 (2), 154–80.

Atkinson, R. and Flint, J (2001) *Accessing Hidden and Hard-to-Reach Populations: Snowball Research Strategies*, University of Surrey.

Australian Bureau of Statistics (1989; 2005) *Overseas Arrivals and Departures*. Canberra, 3014.0.

Axelrod, R. (1997) The dissemination of culture: A model with local convergence and global polarization. *The Journal of Conflict Resolution* 41 (2), 203–26.

Axelrod, R. (2006) Advancing the art of simulation in the social sciences. In J.-P. Rennard (ed.) *Handbook of Research on Nature Inspired Computing for Economy and Management* (Chapter 7). Hersey, PA: Idea Group.

Axelsson, B. and Easton, G. (eds) (1992) *Industrial Networks: A New View of Reality*. London: Routledge.

Baggio, R. (2005) The relationship between virtual and real image of tourism operators. *e–Review of Tourism Research (eRTR)* 3 (5), Retrieved January, 2006, from http://ertr.tamu.edu.

Baggio, R. (2006) Complex systems, information technologies and tourism: A network point of view. *Information Technology and Tourism* 8 (1), 15–29.

Baggio, R. (2007a) The Web Graph of a Tourism System. *Physica A* 379 (2), 727–34.

Baggio, R. (2007b) Symptoms of complexity in a tourism system. *Tourism Analysis* (accepted, preprint at: http://arxiv.org/abs/physics/0701063).

Baggio, R. and Scott, N. (2007) What network analysis of the www can tell us about the organisation of tourism destinations. Paper presented at the CAUTHE, Sydney.

Baggio, R., Antonioli Corigliano, M. and Tallinucci, V. (2007a) The websites of a tourism destination: A network analysis. In M. Sigala, L. Mich and J. Murphy (eds) *Information and Communication Technologies in Tourism 2007 – Proceedings of the International Conference in Ljubljana, Slovenia* (pp. 279–88). Wien: Springer.

Baggio, R., Scott, N. and Wang, Z. (2007b). What network analysis of the WWW can tell us about the organisation of tourism destinations. *Proceedings of the CAUTHE 2007, Sydney, Australia*, 11-14 February.

Balci, O. (2003) Verification, validation, and certification of modeling and simulation applications. *Proceedings of the 2003 Winter Simulation Conference*. Piscataway, NJ, 150–8.

Baloglu, S. and Mangaloglu, M. (2001) Tourism destination images of Turkey, Egypt, Greece, and Italy as perceived by US–based tour operators and travel agents. *Tourism Management* 22 (1), 1–9.

Bankes, S.C. (2002) Tools and techniques for developing policies for complex and uncertain systems. *Proceedings of the National Academy of the Sciences of the USA* 99 (suppl. 3), 7263–6.

Barabási, A.–L. and Bonabeau, E. (2003) Scale–free networks. *Scientific American* 288 (5), 50–9.

Barabási, A.–L. (2002) *Linked: The New Science of Networks*. Cambridge, MA: Perseus.

Barabási, A.–L. and Albert, R. (1999) Emergence of scaling in random networks. *Science* 286, 509–12.

Barrat, A., Barthélemy, M., Pastor–Satorras, R. and Vespignani, A. (2004) The architecture of complex weighted networks. *Proceedings of the National Academy of the Sciences of the USA* 101, 3747–52.

Barrett, C., Eubank, S. and Smith, J. (2005) If smallpox strikes Portland. *Scientific American* 292 (3), 54–61.

Barthélemy, M., Barrat, A., Pastor–Satorras, R. and Vespignani, A. (2005) Characterization and modeling of weighted networks. *Physica A* 346, 34–43.

Barua, A., Whinston, A.B. and Yin, F. (2000) Value and productivity in the internet economy. *Computer* 33 (5), 102–5.

Bar-Yam, Y. (1997) *Dynamics of Complex Systems*. Reading, MA: Addison-Wesley.

Batagelj, V. and Mrvar, A. (1998) Pajek — Program for large network analysis. *Connections* 31 (2), 47–57.

Baum, T. (1998) Taking the exit route: Extending the Tourism Area Life Cycle model. *Current Issues in Tourism* 1 (2), 167–75.

Becken, S. and Gnoth, J. (2004) Tourist consumption systems among overseas visitors: Reporting on American, German, and Australian visitors to New Zealand. *Tourism Management*, 25 (3), 375–85.

Bellamy, C., Horrocks, I. and Webb, J. (1995) Exchanging information with the public: From one–stop shops to community information systems. *Local Government Studies* 21 (1), 11–30.

Bérard, B., Bidoit, M., Finkel, A., Laroussinie, F., Petit, A., Petrucci, L., Schnoebelen, P. and McKenzie, P. (2001) *Systems and Software Verification. Model-checking Techniques and Tools*. Berlin: Springer.

Bernardes, A.T., Stauffer, D. and Kertész, J. (2002) Election results and the Sznajd model on Barabasi network. *The European Physical Journal B* 25, 123–7.

Bernat, E.R., Clukey, T. and Slater, J.R. (2003) Tracking the web visibility of North Country communities. *Adirondack Journal of Environmental Studies* 10 (2), 27–35.

Bertels, K. and Boman, M. (2001) Agent-based social simulation in markets. *Electronic Commerce Research* 1 (1–2), 149–58.

Bezdek, J.C., Keller, J., Krisnapuram, R. and Pal, N.R. (2005) *Fuzzy Models and Algorithms for Pattern Recognition and Image Processing*. New York: Springer Science+Business Media.

Bian, Y. (1994) *Work and Inequality in Urban China*. Albany, N.Y.: State University of New York Press.

Bickerdyke, I. (1996) Australia: The evolving structure and strategies of business networks. In Local Economic and Employment Development (ed.) *Networks of Enterprises and Local Development: Competing and Co–operating in Local Productive Systems* (pp. 203–16). Paris: Organisation for Economic Co–operation and Development.

Biggs, N.L., Lloyd, E.K. and Wilson, R.J. (1976) *Graph Theory 1736–1936*. Oxford, U.K: Clarendon.

Birkin, M., McFarland, O., Dew, P. and Hodrien, J. (2005) HYDRA: A prototype grid-enabled spatial decision support system. *Proceedings of the 1st International Conference on e–Social Science*. Manchester, UK, 22–24 June.

Birnie, S.A. and Horvath, P. (2002) Psychological predictors of Internet social communication. *Journal of Computer–Mediated Communication [On–line]*, 7 (4). Retrieved April, 2006, from http://jcmc.indiana.edu/vol7/issue4/horvath.html.

Bjork, P. and Virtanen, H. (2005) What tourism project managers need to know about co–operation facilitators. *Scandinavian Journal of Hospitality and Touris*, 5 (3), 212–30.

Bjorkman, I. and Kock, S. (1995) Social relationships and business networks: The case of Western companies in China. *International Business* 4, 519–35.

Blackman, C. (2000) *China Business: The Rules of the Game*. St Leonards, NSW: Allen and Unwin.

Blain, C.R. (2001) Destination branding in destination marketing organizations. Unpublished MBA thesis, University of Calgary, Ann Arbor, MI.

Blau, P. (1964). *Exchange and Power in Social Life*. New York: John Wiley.

Bleeke, J. and Ernst, D. (1993) *Collaborating to Compete: Using Strategic Alliances and Acquisitions in the Global Marketplace*. New York: John Wiley and Sons.

Blom–Hansen, J. (1997) A new institutional perspective on policy networks. *Public Administration* 75, 669–93.

Blumberg, K. (2004) Cooperative networks in destination marketing: A case study from Nelson/Tasman region, New Zealand. Paper presented at the Networking and Partnerships in Destination Development and Management Annual Conference, Naples, Italy.

Boccaletti, S., Latora, V., Moreno, Y., Chavez M. and Hwang D.-U. (2006) Complex networks: Structure and dynamics. *Physics Reports* 424 (4–5), 175–308.

Boivin, C.A. (1987) Public–private sector interactions in Canada. In *Travel and Tourism: Thrive or Survive* (pp. 147–50). Seattle, WA: Travel and Tourism Research Association.

Bollobás, B. (1998) *Modern Graph Theory*. New York: Springer.

Bolwijn, P.T. and Kumpe, T. (1990). Manufacturing in the 1990s: Productivity, flexibility and innovation. *Long Range Planning* 4, 44–57.

Bonacich, P. (1972) Factoring and weighting approaches to status scores and clique identification. *Journal of Mathematical Sociology* 2, 113–20.

Booher, D.E., and Innes, J.E. (2002) Network power in collaborative planning. *Journal of Planning Education and Research* 21 (3), 221–36.

Borgatti, S.P., and Foster, P.C. (2003) The network paradigm in organizational research: A review and typology. *Journal of Management* 29 (6), 991–1013.

Borgatti, S.P. (2006) Netdraw, Analytic Technologies, Natick, http://www.analytictech.com/.

Borgatti, S.P., Everett, M.P. and Freeman, L.C. (1999) UCINET 6.0, 1.00 edn, Analytic Technologies, Natick.

Börzel, T.A. (1998) Organizing Babylon: On the different conceptions of policy networks. *Public Administration* 76, 253–73.

Börzel, T.A. and Risse, T. (2005) Public–private partnerships: Effective and legitimate tools of international governance? In E. Grande and L.W. Pauly (eds) *Complex Sovereignty: On the Reconstitution of Political Authority in the 21st Century*. Toronto: University of Toronto Press.

Bott, E. (1957) *Family and Social Network*. London: Tavistock.

Bourdieu, P. (1986) The forms of capital. In J. Richardson (ed.) *Handbook of Theory and Research for the Sociology of Education* (pp. 241–58). New York: Greenwood.

Bramwell, B. (2004) Partnerships, participation and social science research in tourism planning. In A.L Lew, C.M. Hall and A.M. Williams (eds) *A Companion to Tourism* (pp. 541–54). Oxford: Blackwell.

Bramwell, B. and Lane B. (Eds) (2000) *Tourism Collaboration and Partnerships: Politics, Practices and Sustainability*. Clevedon UK: Channel View Publications.

Brandenburger, A.M. and Nalebuff, B.J. (1996) *Co–opetition*. Doubleday, New York

Brandes, U., Kenis, P., Raab, J., Schneider, V. and Wagner, D. (1999) Explorations into the visualization of policy networks. *Journal of Theoretical Politics* 11 (1), 75–106.

Brass, D. and Burkhardt, M. (1992) Centrality and power in organizations. In N. Nohria and R.G. Eccl (eds) *Networks and Organizations: Structure, Form and Action*. Boston: Harvard Business School Press.

Brass, D.J., Butterfield, K.D. and Skaggs, B.C. (1998) Relationships and unethical behavior: A social network perspective. *The Academy of Management Review*, 23, (1) 14–31.

Braun, P. and Hollick, M. (2006) Tourism clusters: Uncovering destination value chains. Paper presented at the 2006 CAUTHE Conference, Melbourne.

Braun, P. (2003a) Networking tourism SMEs: e-commerce and e-marketing issues in regional Australia. *Journal of Information Technology and Tourism* 5 (1), 13–23.

Braun, P. (2003b) Regional tourism networks: The nexus between ICT diffusion and change in Australia. *Journal of Information Technology and Tourism* 6 (4), 231–43.

Breda, Z., Costa, R., and Costa, C. (2006) Do clusters and networks make small places beautiful? The case of Caramulo (Portugal). In L. Lazzeretti and C. Petrillo (eds) *Tourism Local Systems and Networking* (pp. 67–82). Oxford: Elsevier.

Breiger, R.L. (2004) The analysis of social networks In M. Hardy and A. Bryman (eds) *Handbook of Data Analysis*. London: Sage.

Brin, S. and Page, L. (1998) The anatomy of a large–scale hypertextual (Web) search engine. *Computer Networks and ISDN Systems* 30 (1–7), 107–17.

Brinton Milward, H. and Provan, K.G. (1998) Measuring network structure. *Public Administration* (76), 387–407.

Broder, A.Z., Kumar, S.R., Maghoul, F., Raghavan, P., Rajagopalan, S., Stata, R., Tomkins, A. and Wiener, J.L. (2000) Graph structure in the web. *Computer Networks* 33 (1–6), 309–20.

Brunt, P. and Courtney, P. (1999) Host perceptions of sociocultural impacts. *Annals of Tourism Research* 26 (3), 493–515.

Buchanan, M. (2002) *Nexus: Small Worlds and the Ground-breaking Science of Networks*. New York: Norton.

Budowski, G. 1976. Tourism and environmental conservation: Conflict, coexistence, or symbosis. *Environmental Conversation* 3 (1), 27–31.

Buhalis, D. (1998) Strategic use of information technologies in the tourism industry. *Tourism Management* 19 (5), 409–21.

Buhalis, D. (2000) Marketing the competitive destination of the future. *Tourism Management* 21 (1), 97–116.

Buhalis, D. and Molinaroli, E. (2003) Entrepreneurial networks and supply communities in the Italian eTourism. *Information Technology and Tourism* 5 (3), 175–85.

Burstein, P. (1991) Policy domains: Organization, culture, and policy outcomes. *Annual Review of Sociology* 17, 327–50.

Burt, R.S. (1987). Social contagion and innovation: Cohesion versus structural equivalence. *American Journal of Sociology*, 92 (6), 1287–1335.

Burt, R.S. (1990) *Political Networks: The Structural Perspective*. Cambridge: Cambridge University Press.

Burt, R.S. (1992) The social structure of competition. In N. Nohria and R. G. Eccles (eds) *Networks and Organizations: Structure, Form and Action*. Boston: Harvard Business School Press.

Burt, R.S. (1997) A note on social capital and network content. *Social Networks* 19, 355–73.

Burt, R.S. (2000) The network structure of social capital. In R. I. Sutton and B. M. Staw (eds) *Research in Organizational Behavior 22* (pp. 345–423). Greenwich, CT: JAI Press.

Burt, R.S. (1980a) Innovation as a structural interest: Rethinking the impact of network position on innovation adoption. *Social Networks* 2, 327–55.

Burt, R.S. (1980b) Co-optive corporate actor networks: A reconsideration of interlocking directorates involving American manufacturing. *Administrative Science Quarterly* 25 (4), 557–82.

Burt, R.L. and Minor, M. (eds) (1983) *Applied Network Analysis*. London: Sage Publications.

Butler, J. (1990) *Gender Trouble*. London: Routledge.

Butler, R. W. (ed.) (2005a) *The Tourism Area Life Cycle, Vol. 1: Applications and Modifications*. Clevedon: Channel View.

Butler, R. W. (ed.) (2005b) *The Tourism Area Life Cycle, Vol. 2: Conceptual and Theoretical Issues*. Clevedon: Channel View.

Caffyn, A. and Jobbins, G. (2003) Governance capacity and stakeholder interactions in the development and management of coastal tourism: Examples from Morocco and Tunisia. *Journal of Sustainable Tourism* 11 (2–3), 224–45.

Cai, L. (2002) Cooperative branding for rural destinations. *Annals of Tourism Research* 29 (3), 720–42.

Calantone, R.J. and Mazanec, J.A. (1991) Marketing management and tourism. *Annals of Tourism Research* 18 (1), 101–19.

Callaway, D.S., Newman, M.E.J., Strogatz, S.H. and Watts, D.J. (2000) Network robustness and fragility: Percolation on random graphs. *Physical Review Letters* 85, 5468–71.

Carlson, J. (1999) A systems approach to island tourism destination management. *Systems Research and Behavioral Science*, 16 (4), 321–7.

Castells, M. (1997) *The Power of Identity*. Malden, MA: Blackwell.

Cegarra-Navarro, J.G. (2005) An empirical investigation of organizational learning through strategic alliances between SMEs. *Journal of Strategic Marketing* 13, 3–13.

Chamberlain, J. (1992) On the tourism trail: A nice little earner, but what about the cost? *North and South* September, 88–97.

Chen, M. (1995) *Asian Management System: Chinese Japanese and Korean Styles*. London and New York: Routledge.

Cheong, S.M. and Miller, M.L. (2000) Power and tourism: A Foucauldian observation. *Annals of Tourism Research* 27 (2), 371–90.

Chetty, S., Eriksson, K. and Hohental, J. (2003) Collaborative experience in internationalising firms. In A. Blomstermo and D. Sharma (eds) *Learning in the Internationalisation Process of Firms* (pp. 56–73). Cheltenham: Edward Elgar.

Chiesi, A.M. (2001) Network analysis. In N.J.Smesler and P.B. Bates (eds) *International Encyclopedia of the Social and Behavioral Sciences* (pp. 10499–10502). Oxford: Elsevier Science.

Chinese Contemporary Dictionary (1983). Beijing: Commercial Publication (Shang Wu Yin Shu Guan) (in Chinese).

Chua, A. (2001) Relationship between the types of knowledge shared and types of communication channels used. *Journal of Knowledge Management Practice* 2, (http:www.tlainc.com/jkmp2.htm accessed 15/3/04).

Clegg, S. (1989) *Frameworks of Power*. London: Sage.

Cochran, W.G. (1977) *Sampling Techniques* (3rd edn). New York: John Wiley.

Coleman, J. (1988) Social capital in the creation of human capital. *American Journal of Sociology* 94 (Supplement), S95–S120.

Commonwealth Government (Australia) (1983) Review of Commonwealth Administration: Report January 1983. Canberra: AGPS.

Conte, D., Foggia, P., Sansone, C. and Vento, M. (2004) Thirty years of graph matching in pattern recognition. *International Journal of Pattern Recognition and Artificial Intelligence* 18 (3), 265–98.

Conte, R., Hegselmann, R. and Terna, P. (1997) *Simulating Social Phenomena*. Berlin: Springer-Verlag.

Conway, R.W. (1963) Some tactical problems in digital simulation. *Management Science* 10 (1), 47–61.

Conway, R.W., Johnson, B.M. and Maxwell, W.C. (1959) Some problems of digital systems. *Management Science* 6 (1), 92–110.

Cook, K.S., and Whitmeyer, J.M. (1992) Two approaches to social structure: Exchange theory and social structure. *Annual Review of Sociology* 18, 109–27.

Cooper, C. (2006) Knowledge management and tourism. *Annals of Tourism Research* 33 (1), 47–64

Cooper, C. and Scott, N. (2005) Knowledge for networked destinations. Paper presented at the Recent Developments in Tourism Research Conference, Faro, Portugal.

Costa, C. (1996) Towards the improvement of the efficiency and effectiveness of tourism planning and development at the regional level: Planning and networks. The case of Portugal. Unpublished PhD Thesis. Guildford: University of Surrey.

Costa, J. and Eccles, G. (1996) Hospitality and tourism impacts: An industry perspective. *International Journal of Contemporary Hospitality Management* 8 (7), 11.

Costa, R. (2005) Avaliação do Potencial de Crescimento e Desenvolvimento das Pequenas e Micro Empresas do Sector do Turismo. Unpublished MSc Thesis. Aveiro: Universidade de Aveiro.

Costenbader, E. and Valente, T.W. (2003) The stability of centrality measures when networks are sampled. *Social Networks* 25, 283–307.

Coviello, N. and Munro, H. (1995) Growing the entrepreneurial firm: Networking for international market development. *European Journal of Marketing* 29 (7), 49–61.

Coviello, N. and Munro, H. (1997) Network relationships and the internationalisation process of the small software firm. *International Business Review* 6 (4), 361–86.

Cravens, D.W. and Piercy, N.F. (1994) Relationship marketing and collaborative networks in service organizations. *International Journal of Service Industry Management* 5 (5), 39–53.

Cropper, S. (1996) Collaborative working and the issue of sustainability. In C. Huxham (ed.) *Creating Collaborative Advantage* (pp. 80–100). London: Sage Publications.

Cross, R., Borgatti, S.P. and Parker, A (2002) Making invisible work visible: Using social network analysis to support strategic collaboration. *California Management Review* 44 (2), 25–46.

Crotts, J.C., Buhalis, D. and March, R. (2000) Introduction: Global alliances in tourism and hospitality management. In J. C. Crotts, D. Buhalis and R. March (eds) *Global Alliances in Tourism and Hospitality Management* (pp. 1–10). Binghamton, NY: The Haworth Hospitality Press.

Crucitti, P., Latora, V., Marchiori, M. and Rapisarda, A. (2003) Efficiency of scale-free networks: Error and attack tolerance. *Physica A* 320, 622–42.

Crucitti, P., Latora, V., Marchiori, M. and Rapisarda, A. (2004) Error and attack tolerance of complex networks. *Physica A* 340, 388–94.

Crutchfield, J.P. and Young, K. (1990) Computation at the onset of chaos. In W. Zurek (ed.) *Entropy, Complexity, and the Physics of Information [SFI Studies in the Sciences of Complexity]* 8, 223–69. Reading, MA: Addison-Wesley.

Csermely, P. (2004) Strong links are important, but weak links stabilize them. *TRENDS in Biochemical Sciences* 29 (7), 331–4.

Curran, J., Jarvis, R., Blackburn, R.A. and Black, S. (1993) Networks and small firms: Constructs, methodological strategies and some findings. *International Small Business Journal* 11 (2), 13–25.

Da Costa, L. and Terhesiu, D. (2005) A simple model for the diffusion of ideas (Complex Systems Summer School Final Project Papers) Santa Fe, NM: Santa Fe Institute. Retrieved September, 2006, from http://www.santafe.edu/education/csss/csss05/papers/.

da F. Costa, L., Rodrigues, A., Travieso, G. and Villas Boas, P.R. (2005) Characterization of complex networks: A survey of measurements (Preprint arxiv/cond–mat/0505185): ArXiv e–prints archive. Retrieved June, 2006, from http://arxiv.org/abs/cond–mat/0505185.

Dahl, R.A. (1961) *Who Governs? Democracy and Power in an American City*. New Haven: Yale University Press.

Dann, G. (1996) *The Language of Tourism: A Sociolinguistic Perspective*. Wallingford, England: CAB International.

Davenport, T. and L. Prusak (1998) *Working Knowledge: How Organizations Manage What They Know*. Boston: Harvard Business School Press.

Davies, B. (2003) The role of quantitative and qualitative research in industrial studies of tourism. *International Journal of Tourism Research* 5, 97–111.

Davies, H., Leung, T., Luk, S. and Wong, Y. (1995) The benefits of 'Guanxi': The value of relationships in developing the Chinese market. *Industrial Marketing Management* 24, 207–14.

Davis, G. and Weller, P. (eds) (2000) *Are You Being Served? State, Citizens and Governance*. Sydney: Allen and Unwin.

de Araujo, L.M., and Bramwell, B. (2000) Stakeholder assessment and collaborative tourism planning: The case of Brazil's Costa Dourada project. In B. Bramwell and B. Lane (eds) *Tourism, Collaboration, and Partnerships: Politics, Practice, and Sustainability* (pp. 272–94). Clevedon: Channel View.

de Chernatony, L. and Harris, F. (2000) Developing corporate brands through considering internal and external stakeholders. *Corporate Reputation Review* 3 (3), 268–74.

de Nooy, W., Mrvar, A. and Batagelj, V. (2005) *Exploratory Social Network Analysis with Pajek*. Cambridge: Cambridge University Press.

Deo, N. and Gupta, P. (2001) Graph–theoretic web algorithms: An overview. In T. Böhme and H. Unger (eds) *Lecture Notes in Computer Science Vol. 2026* (pp. 91–102). Berlin: Springer.

Department of Tourism, Industry and Resources (2005) *National Tourism Emerging Markets Strategy: China and India, Investing Today for Tomorrow*, Canberra.

Deslandes, D.D. (2003) Assessing consumer perceptions of destinations: A necessary first step in the destination branding process. Unpublished PhD Thesis, Florida State University, Ann Arbor, MI.

Detlor, B., Sproule, S. and Gupta, C. (2003) Pre-purchase online information seeking: Search versus browse. *Journal of Electronic Commerce Research* 4 (2), 72–84.

Dhyani, D., Ng, W.K. and Bhowmick, S.S. (2002) A survey of web metrics. *ACM Computing Surveys* 34 (4), 469–503.

Dick, P. and Cassell, C. (2004) The position of policewomen: A discourse analytic study. *Work, Employment and Society* 18 (1), 51–72.

Diestel, R. (2005) *Graph Theory*, Electronic edition 2005. New York, NY: Springer.Online at http://www.math.uni–hamburg.de/home/diestel/books/graph.theory/GraphTheoryIII.pdf [last access: January 2006].

Dill, S., Kumar, S. R., McCurley, K., Rajagopalan, S., Sivakumar, D. and Tomkins, A. (2001) Self similarity in the web. *Proceedings of the 27ᵗʰ International Conference on Very Large Data Bases, Rome, Italy* (September 11–14), 69–78.

Dill, S., Kumar, S. R., McCurley, K., Rajagopalan, S., Sivakumar, D. and Tomkins, A. (2002) Self similarity in the web. *ACM Transactions on Internet Technology (TOIT)* 2 (3–August), 205–23.

DiMaggio, P.J. and Powell, W.W. (1983) The iron cage revisited: Institutional isomorphism and collective rationality in organisational fields. *American Sociological Review* 48, 147–60.

Dinnie, K. (2002) Implications of national identity for marketing strategy. *Marketing Review* 2 (3), 285–300.

Dollinger, M.J. (1990) The evolution of collective strategies in fragmented industries. *Academy of Management Review* 15 (2), 266–85.

Dongarra, J. and Sullivan, F. (2000) Guest Editors' Introduction: The Top 10 Algorithms. *Computing in Science and Engineering* 2 (1), 22–3.

Doreian, P. and Stokman, F.N. (1997) *Evolution of Social Networks.* Amsterdam: Gordon and Breach Publishers.

Dorogovtsev, S.N. and Mendes, J.F.F. (2002) Evolution of networks. *Advances in Physics* 51, 1079–187.

Dorogovtsev, S.N. and Mendes, J.F.F. (2003) *Evolution of Networks: From Biological Nets to the Internet and WWW.* Oxford: Oxford University Press.

Dorogovtsev, S.N., Mendes, J.F.F. and Samukhin, A.N. (2000) Structure of growing networks with preferential linking. *Physical Review Letters* 85, 4633–6.

Dowding, K. (1995) Model or metaphor? A critical review of the policy network approach. *Political Studies* 43 (2), 136–58

Dowding, K. (2001) There must be end to confusion: Policy networks, intellectual fatigue, and the need for political science methods courses in British universities. *Political Studies* 49, 89–105.

Dredge, D. (2004) Networks, conflict and collaboration: Tourism planning. *Proceedings of the Creating Tourism Knowledge – CAUTHE Conference, Brisbane, February 10–15*, 195–207.

Dredge, D. (2005) Networks and innovation in Lake Macquarie. In D. Carson and J. Macbeth (eds) *Regional Tourism Cases: Innovation in Regional Tourism* (pp. 61–68). Gold Coast: STCRC.

Dredge, D. (2006a) Policy networks and the local organisation of tourism. *Tourism Management* 27 (2), 269–80.

Dredge, D. (2006b) Networks, conflict and collaborative communities. *Journal of Sustainable Tourism.* 14 (6), 562–81.

Dredge, D., and Jenkins, J.M. (2003) Destination place identity and regional tourism policy. *Tourism Geographies* 5 (4), 383–407.

Dredge, D. and Jenkins, J.M. (eds) (2007) *Tourism Policy and Planning.* Brisbane: John Wiley and Sons.

Dunfee, T.W. and Warren, D.E. (2001) Is guanxi ethical? A normative analysis of doing business in China. *Journal of Business Ethics* 32, 191–204.

Dyer, J. and Nobeoka, K. (2000) Creating and managing a high-performance knowledge-sharing network: The Toyota case. *Strategic Management Journal* 21, 345–67.

Eades, P (1984) A heuristic for graph drawing. *Congressus Numerantium* 42, 149–60.

East Asia Analytical Unit (1995) Overseas Chinese business networks in Asia. Canberra: East Asia Analytical Unit, Department of Foreign Affair and Trade.

Echtner, C.M. and Jamal, T.B. (1997) The disciplinary dilemma of tourism studies. *Annals of Tourism Research* 24 (4), 868–83.

Edmonds, B. (2004) Against the inappropriate use of numerical representation in social simulation (Technical Report CPM–04–129): Centre for Policy Modelling, Manchester Metropolitan University. Retrieved November, 2006, from http://cogprints.org/4262/.

Edmonds, B. and Chattoe, E. (2005) When simple measures fail: Characterising social networks using simulation. Paper presented at the Social Network Analysis: Advances and Empirical Applications Forum, Oxford (July 16–17).

Efe, K., Raghavan, V., Chu, H., Broadwater, A., Bolelli, L. and Ertekin, S. (2000) The shape of the web and its implications for searching the web. *Proceedings of the International Conference on Advances in Infrastructure for Electronic Business, Science, and Education on the Internet (SSGRR'00), L'Aquila, Italy*, 31–6.

Efron, B. and Tibshirani, R.J. (1993) *An Introduction to the Bootstrap*. New York: Chapman and Hall.

Eisenhardt, K. and Martin, J. (2000) Dynamic capabilities: What are they? *Strategic Management Journal* 21, 1105–21.

Erdős, P. and Rényi, A. (1959) On random graphs. *Publicationes Mathematicae (Debrecen)* 6, 290–7.

Erdős, P. and Rényi, A. (1960) On the evolution of random graphs. *Publications of the Mathematical Institute of the Hungarian Academy of Sciences* 5, 17–61.

Erdős, P. and Rényi, A. (1961) On the strength of connectedness of a random graph. *Acta Mathematica Academiae Scientiarum Hungaricae* 12, 261–7.

Etamad, H. and Wright, R. (2003) Globalization and entrepreneurship. In H. Etamad and R. Wright (eds) *Globalization and Entrepreneurship: Policy and Strategy Perspectives* (pp. 3–14). Cheltenham: Edward Elgar.

Etemad, H. (2004) *International Entrepreneurship in Small and Medium Size Enterprises: Orientation, Environment and Strategy*. Northampton: Edward Elgar.

Euler, L. (1736) Solutio problematis ad geometriam situs pertinentis. *Commentarii Academiae Scientiarum Imperialis Petropolitanae* 8, 128–40.

Faloutsos, M., Faloutsos, P. and Faloutsos, C. (1999) On power-law relationships of the internet topology. *Computer Communication Review* 29, 251–62.

Farrell, B.H. and Twining-Ward, L. (2004) Reconceptualizing tourism. *Annals of Tourism Research* 31 (2), 274–95.

Farrell, M. and Oczkowski, E. (2002) Are market orientation and learning orientation necessary for superior organizational performance? *Journal of Market-Focused Management* 5 (3), 197–217.

Faulkner, B. (2000) The future ain't what it used to be: Coping with change, turbulence and disasters in tourism research and destination management (Professorial Lecture Series No. 6, 2000): Griffith University. Retrieved July, 2005, from http://www.gu.edu.au/ins/collections/proflects/faulkner00.pdf.

Faulkner, B. and Russell, R. (1997) Chaos and complexity in tourism: In search of a new perspective. *Pacific Tourism Review* 1, 93–102.

Faulkner, B. and Russell, R. (2001) Turbulence, chaos and complexity in tourism systems: A research direction for the new millennium. In B. Faulkner, G. Moscardo and E. Laws (eds) *Tourism in the 21st Century: Lessons from Experience* (pp. 328–349). London: Continuum.

Fennell, D.A. and Butler, R.W. (2003) A human ecological approach to tourism interactions. *The International Journal of Tourism Research* May/Jun, 5 (3), 197–210.

Fesenmaier, D.J. and Parks, D. (1998) ITMES: A knowledge-based system for the tourism industry. In D. Buhalis, A. Min Toja and J. Jafari (eds) *Information and Communication Technologies in Tourism* (pp. 1–5). Vienna: Springer Verlag.

Fesenmaier, D.R., and Uysal, M. (1993) *Communication and Channel Systems in Tourism Marketing*. New York: Haworth Press.

Flake, G.W. (1998) *The Computational Beauty of Nature*. Cambridge, MA: MIT Press.

Flake, G.W., Lawrence, S., Giles, C.L. and Coetzee, F.M. (2002) Self-organization of the Web and identification of communities. *IEEE Computer* 35 (3), 66–71.

Florida, R. (1995) Toward the learning region. *Futures* 27, 527–36.

Foley, A., and Fahy, J. (2004) Incongruity between expression and experience: The role of imagery in supporting the positioning of a tourism destination brand. *Journal of Brand Management* 11(3), 209–17.

Ford, D., Gadde, L.E., Hakansson, H., Lundgren, A., Snehota, I., Turnbull, P. and Wilson, D. (1998) *Managing Business Relationships*. Chichester: John Wiley and Sons.

Forsgren, M., Holm, U. and Johanson, J. (2005) *Managing the Embedded Multinational: A Business Network View*. Cheltenham: Edward Elgar.

Foucault, M. (1973) *The Birth of the Clinic: An Archaeology of Medical Perception*. London: Tavistock.

Foucault, M. (1979) *Discipline and Punish: The Birth of the Prison*. Harmondsworth, England: Penguin.

Foucault, M. (1980) Power, right, truth. In C. Gordon (ed.) *Power/Knowledge: Selected Interviews and Other Writings, 1972–1977* (pp. 92–108) New York: Pantheon Books.

Fox, N.J. (1988) Foucault, Foucauldians and sociology. *The British Journal of Sociology* 49 (3), 415–33.

Frank, O. (1971) *Statistical Inference in Graphs*. Stockholm: Försvarets forskningsanstalt (Research Institute of National Defense).

Fredline, E. and Faulkner, B. (2000) Host community reactions: A cluster analysis. *Annals of Tourism Research* 27 (3), 763–84.

Freeman, L.C. (2004) *The Development of Social Network Analysis: A Study in the Sociology of Science*. Vancouver, North Charleston: Empirical Press.

Freeman, R. (1984) *Strategic Management: A Stakeholder Approach*. Boston: Pitman.

French, C.N., Collier, A. and Craig-Smith, S.J. (2000) *Principles of Tourism*. Sydney: Longman.

Frew, A.J. (2000) Information and communications technology research in the travel and tourism domain: Perspective and direction. *Journal of Travel Research* 39, 136–45.

Friedman, A.L. and Miles, S. (2002) SMEs and the environment: Evaluating dissemination routes and handholding levels. *Business Strategy and the Environment* 11, 324–41.

Frost, W. (2004) Tourism, rainforests and worthless lands: The origins of National Parks in Queensland. *Tourism Geographies* 6(4), 493–507.

Fukugawa, N. (2006). Determining factors in innovation of small-firm networks: A case of cross-industry groups in Japan. *Small Business Economics* 27(2), 181–93.

Fulop, L. and Richards, D. (2002) Connections, culture and context: Business relationships and networks in the Asia–Pacific Region. In C. Harvie and B.C. Lee (eds) *Globalisation and SMEs in East Asia: Studies of Small and Medium Sized Enterprises in East Asia*. Cheltenham: Edward Elgar.

Fyall, A. and Garrod, B. (2005) *Tourism Marketing: A Collaborative Approach.* Clevedon: Channel View.

Galaskiewicz, J. and Wasserman, S. (1993) Social network analysis: Concepts, methodology, and directions for the 1990s. *Sociological Methods and Research* 22 (1), 3–22.

Galaskiewicz, J. (1996) The 'New network analysis' and its application to organizational theory and behaviour. In D Iacobucci (ed.) *Networks in Marketing* (pp. 19–31). Thousand Oaks, CA: Sage Publications.

Galaskiewicz, J. and Burt, R.S. (1991) Interorganisation contagion in corporate philanthropy. *Administrative Science Quarterly*, March, 88–105.

Galaskiewicz, J. and Wasserman, S. (1989) Mimetic and normative processes within an interorganizational field: An empirical test. *Administrative Science Quarterly* 34, 454–80.

Galbraith, J.K. (1983) *The Anatomy of Power.* Boston: Houghton Mifflin.

Gamm, L. (1981) An introduction to research in interorganizational relations. *Journal of Voluntary Action Research* 10 (1), 18–52.

Gentle, J.E., Härdle, W. and Mori, Y. (eds) (2004) *Handbook of Computational Statistics: Concepts and Methods.* Heidelberg: Springer.

Getz, D. and Jamal, T. (1994) The environment–community symbiosis: A case for collaborative tourism planning. *Journal of Sustainable Tourism* 2 (3), 152–73.

Gibson, D., Kleinberg, J. and Raghavan (1998) Inferring Web communities from link topology. *Proceedings of the 9th ACM Conference on Hypertext and Hypermedia* 225–34.

Gibson, L., Lynch, P. and Morrison, A. (2005) The local destination tourism network: Development issues. *Tourism and Hospitality Planning and Development* 2 (2), 87–99.

Giddens, A. (1993) *Sociology* (2nd rev. edn). Cambridge: Polity Press.

Giddens, A. (1998) *The Third Way: The Renewal of Democracy.* Oxford: Policy Press.

Gilbert, N. (1999) Simulation: A new way of doing social science. *American Behavioral Scientist* 42, 1485–7.

Gilchrist, A. (2004) *The Well-Connected Community: A Networking Approach to Community Development.* Bristol: Policy Press.

Gleick, J. (1987) *Chaos: Making a New Science.* New York: Viking.

Glover, T. and Hemingway, J. (2005) Locating leisure in the social capital literature. *Journal of Leisure Research* 37 (4), 387–401.

Godsyl, C. and Royle, G. (2001) *Algebraic Graph Theory.* New York: Springer.

Grabher, G. (2006) Trading routes, bypasses, and risky intersections: Mapping the travels of 'networks' between economic sociology and economic geography. *Progress in Human Geography* 30 (2), 163–89.

Grangsjo, Y.v.F. (2003) Destination networking: Co-opetition in peripheral surroundings. *International Journal of Physical Distributionand Logistics Management* 33 (5), 427–48.

Grangsjo, Y.v.F. (2006) Hotel networks and social capital in destination marketing. *International Journal of Service Industry Management* 17 (1), 58–75.

Granovetter, M. (1973) The strength of weak ties. *American Journal of Sociology* 78 (6), 1360–80.

Granovetter, M. (1976). Network sampling: Some first steps. *The American Journal of Sociology* 81 (6), 1287–1303.

Granovetter, M. (1985) Economic action and social structure: The problem of embeddedness. *American Journal of Sociology* 91 (3), 481–510.

Granovetter, M. S. (1992) Problems of explanation in economic sociology. In N. Nohria and R.G. Eccles (eds) *Networks and Organizations: Structure, Form and Action*. Boston: Harvard Business School Press.

Gray, B. (1989) *Collaborating: Finding Common Ground for Multiparty Problems*. San Francisco: Jossey-Bass.

Greffe, X. (1994) Is rural tourism a lever for economic and social development? In B. Bramwell and B. Lane (eds) *Rural Tourism and Sustainable Rural Development*. Clevedon: Channel View.

Gretzel, U. and Fesenmaier, D.R. (2002) Implementing knowledge-based interfirm networks in heterogeneous B2B environments: A case study of the Illinois tourism network. In K.W. Wober, A.J. Frew and M. Hitz (eds) *Information and Communication Technologies in Tourism* (pp. 39–48). Vienna: Springer Verlag.

Guillaume, J.-L. and Latapy, M. (2002) The Web graph: An overview. *Proceedings of the ALGOTEL 2002, Quatrièmes rencontres francophones sur les aspects algorithmiques des télécommunications, Mèze, France*, 52–61.

Gulati, R. (1998) Alliances and networks. *Strategic Management Journal* 19, 293–317.

Gulati, R. and Gargiulo, M. (1999) Where do interorganizational networks come from? *American Journal of Sociology* 104 (5), 1439–93.

Gunn, C.A. (1988a) *Tourism Planning*. New York: Taylor and Francis.

Gunn, C.A. (1988b) *Vacationscape: Designing Tourist Regions* (2nd edn). New York: Van Nostrand Reinhold.

Gunn, C.A. and Var, T. (2002) *Tourism Planning: Basics, Concepts, Cases* (4th edn). New York: Routledge.

Guthrie, D. (1998) The declining significance of guanxi in China's economic transition. *The China Quarterly* 154, 254–82.

Hackathorn, R. (2003) The link is the king. *DM Review [On–line]* August. Retrieved December, 2003, from http://www.dmreview.com.

Haga, T. (2005) Action research and innovation in networks, dilemmas and challenges: Two cases. *AI and Society* 19 (4), 362–83.

Hakansson, H. and Johanson, J. (1992) A model of industrial networks. In B. Axelsson and G. Easton (eds) *Industrial Networks: A New View of Reality* (pp. 28–34). London: Routledge.

Haley, G.T. and Tan, C.T. (1999) East vs. West: Strategic marketing management meets the Asian networks. *Journal of Business and Industrial Marketing* 14, 91–101.

Hall, C.M. (1994) *Tourism and Politics: Policy, Power and Place*. New York: Wiley.

Hall, C.M. (1995) *Introduction to Tourism in Australia: Impacts, Planning and Development*. Australia: Longman.

Hall, C.M. (1999) Rethinking collaboration and partnership: A public policy perspective. *Journal of Sustainable Tourism* 7 (3/4), 274–89.

Hall, C.M. (2000) *Tourism Planning: Policies, Processes and Relationships*. Harlow: Prentice Hall.

Hall, C.M. (2004) Small firms and wine and food tourism in New Zealand: Issues of collaboration, clusters and lifestyles. In R. Thomas (ed.) *Small Firms in Tourism: International Perspectives* (167–81). London: Elsevier.

Hall, C. and Jenkins, J. (1995) *Tourism and Public Policy*. London: Routledge.

Hall, E.T. (1976) *Beyond Culture*. New York: Anchor Press-Doubleday.

Halme, M. and Fadeeva, Z. (2000) Small and medium-sized tourism enterprises in sustainable development networks. *Greener Management International* 30 (Summer), 97–113

Halme, M. (2001) Learning for sustainable development in tourism networks. *Business Strategy and the Environment* 10, 100–14.

Halpin, B. (1999) Simulation in sociology. *American Behavioral Scientist* 42, 1488–1508.

Han, J.J., Dupuy, D., Bertin, N., Cusick, M.E. and Vidal, M. (2005) Effect of sampling on topology predictions of protein–protein interaction networks. *Nature Biotechnology* 23 (7), 839–44.

Hanneman, R. and Riddle, M. (2005) *Introduction to Social Network Methods*. Riverside, CA (published in digital form at http://faculty.ucr.edu/~hanneman/): University of California.

Harré, R. and Madden, E.H. (1975) *Causal Powers: A Theory of Natural Necessity*. Totowa, N.J.: Rowman and Littlefield.

Harris, P.R. and Moran, R.T. (2000) *Managing Cultural Differences: Leadership Strategies for a New World of Business*. Houston, Texas: Gulf Publishing.

Harrison, D. (2004) Tourism in Pacific Islands. *The Journal of Pacific Studies* 26 (1and2), 1–28.

Hassid, J. (2003) *Internationalisation and Changing Skill Needs in European Small Firms: The Services Sector*. Luxembourg: Cedefop.

Hawkinson, J. and Bates, T. (1996) Guidelines for creation, selection and registration of an autonomous system (AS) (RFC 1930): *Internet Engineering Task Force*. Retrieved October, 2006, from http://www.ietf.org/rfc/rfc1930.txt.

Haythornthwaite, C (1996) Social network analysis: An approach and technique for the study of information exchange. *Library and Information Science Research* 18 (4), 323–42.

Healey, P. (1997) *Collaborative Planning: Shaping Places in Fragmented Societies*. London: Macmillan.

Healey, P. (2003) Collaborative planning in perspective. *Planning Theory* 2 (2), 101–23.

Heclo, H. (1978) Issue networks and the executive establishment. In A. King (ed.) *The New American Political System* (pp. 87–124). Washington: American Enterprise Institute.

Held, D. (1989) *Political Theory and the Modern State: Essays on State, Power and Democracy*. Oxford: Policy Press.

Henrickson, L. and McKelvey, B. (2002) Foundations of 'new' social science: Institutional legitimacy from philosophy, complexity science, postmodernism, and agent-based modeling. *Proceedings of the National Academy of the Sciences of the USA* 99 (suppl. 3), 7288–95.

Héritier, A. (1993) Policy-Analyse. Elemente der Kritik und Perspektiven der Neuorientierung. In A. Héritier (Hrsg.) *Policy Analyse. Kritik und Neuorientierung. Politische Vierteljahresschrift, Sonderheft* 24 (pp. 9–36). Opladen: Westdeutscher Verlag.

Hirst, P. (2000) Democracy and governance. In J. Pierre (ed.) *Debating Governance*. Oxford: Oxford University Press.

Hislop, D. (2005) The effect of network size on intra-network knowledge processes. *Knowledge Management Research and Practice* 3 (4), 244.

Hislop, D., Newell, S., Scarbrough, H. and Swan, J. (1997) Innovation and networks: Linking diffusion and implementation. *International Journal of Innovation Management* 1 (4), 427–48.

Hjalager, A.M. (2000) Tourism destinations and the concept of industrial districts. *Tourism and Hospitality Research* 2 (3).

Hjalager, A.M. (2002) Repairing innovation defectiveness in tourism. *Tourism Management* 23, 465–74.

Hogwood, B. and Gunn, L. (1984) *Policy Analysis for the Real World*. New York: Oxford University Press.

Holbrook, J.A. and Wolfe, D.A. (2005) The innovation systems research network: A Canadian experiment in knowledge management. *Science and Public Policy* 32 (2), 109–18.

Holmlund, M. and Kock, S. (1998) Relationships and the internationalism of Finnish small and medium-sized companies. *International Small Business Journal* 16, 46–63.

Homans, C.G. (1958) Social behaviour as exchange. *American Journal of Sociology* 62, 597–606.

Homans, C.G. (1974) *Social Behaviour* (2nd edn). New York: Harcourt-Brace.

Howard, R. (1990) Can small business help countries compete? *Harvard Business Review* November-December, 88–103.

Howlett, M. and Ramesh, M. (1995) *Studying Public Policy: Policy Cycles and Policy Subsystems*. Toronto: Oxford University Press.

Hughes, O.E. (2003) *Public Management and Administration* (3rd edn). Basingstoke: Palgrave.

Huson, M. and Nanda, D. (1995) The impact of Just-In-Time manufacturing on firm performance in the US. *Journal of Operations Management* 12 (3), 297–310.

Hutchings, K. (2002) Improving selection processes but providing marginal support: A review of cross-cultural difficulties for expatriates in Australian organisations in China. *Cross Cultural Management* 9, 32–57.

Huxham, C. (1993) Pursuing collaborative advantage. *The Journal of the Operational Research Society* 44 (6), 599–611.

Huxham, C. (1996) *Creating Collaborative Advantage*. London: Sage.

Iacobucci, D. and Hopkins, N. (1992) Modelling dyadic interactions and networks in marketing. *Journal of Marketing Research* 29 (1), 5–17.

Im, H.J. (2003) An exploratory study of destination branding for the State of Oklahoma. Unpublished PhD thesis. Ann Arbor, MI: Oklahoma State University.

Immergut, E. (1998) The theoretical core of the new institutionalism. *Politics and Society* 26 (1), 5–34.

Inbar, M. and Stoll, C.S. (1972) *Simulation and Gaming in Social Science*. New York: Free Press.

Ingram, P. and Roberts, P.W. (2000) Friendships among competitors in the Sydney hotel industry. *American Journal of Sociology* 106 (2), 387–423.

IWSTAT (2006) Internet Usage Statistics – The Big Picture. Internet World Stats, Retrieved November 2006 from http://www.internetworldstats.com/.

Jackson, M.H. (1997) Assessing the structure of communication on the world wide web. *Journal of Computer-Mediated Communication [On–line]* 3 (1) Retrieved July 2006 from http://jcmc.indiana.edu/vol3/issue1/jackson.html.

Jafari, J. and Ritchie, B.J.R. (1981) Toward a framework for tourism education: Problems and prospects. *Annals of Tourism Research* 8 (1), 13–34.

Jaffe, A. (1989) Real effects of academic research. *American Economic Review* 79 (5), 957–70.

Jamal, T. and Getz, D. (1995) Collaboration theory and community tourism planning. *Annals of Tourism Research* 22 (1), 186–204.

Jarillo, J.C. (1988) On strategic networks. *Strategic Management Journal* 9 (1), 31–41.

Jarillo, J.C. (1993) *Strategic Networks: Creating the Borderless Organisation.* Oxford: Butterworth-Heinemann.

Jary, D. (2002) The global Third Way debate. *The Sociological Review,* 50 (3), 437–49.

Jayawardena, C. and Ramajeesingh, D. (2003) Performance of tourism analysis: A Caribbean perspective. *International Journal of Contemporary Hospitality Management* 15 (3), 176.

Jenkins, J. and Hall, M.C. (1997) Tourism planning and policy in Australia. In M.C. Hall, J. Jenkins and G. Kearsley (eds) *Tourism Planning and Policy in Australia* (pp. 37–48). Sydney: Irwin Publishers.

Jenkins, J.M. (2000) The dynamics of regional tourism organisations in New South Wales, Australia: History, structures and operations. *Current Issues in Tourism* 3 (3), 175–203.

Jenkins, J.M. (2001) Statutory authorities in whose interests? The case of Tourism New South Wales, the Bed Tax, and 'The Games'. *Pacific Tourism Review* 4, 201–18.

Jin, E.M., Girvan, M. and Newman, M.E.J. (2001) The structure of growing social networks. *Physical Review E* 64, 046132.

Johanson, J. and Vahlne, J.E. (1990) The mechanism of internationalisation. *International Marketing Review* 7 (4), 11–24.

Johnson, D.J. (1996) *Information Seeking: An Organisational Dilemma.* Westport: Quorum Books.

Jones, C., Hesterly, W. and Borgatti, S. (1997) A general theory of network governance: Exchange conditions and social mechanisms. *Academy of Management Review* 22 (4), 911–45.

Jones, S. (2005). Community-based ecotourism: The significance of social capital. *Annals of Tourism Research* 32 (2), 303–24.

Joppe, M. (1996) Sustainable community tourism revisited. *Tourism Management* 17 (7), 475–9.

Kaltenborn, B.P. (1996) Keeping tourism under ecological limits. *Ecodecision* 20, 25–8.

Kaplanidou, K. and Vogt, C. (2003) *Destination Branding: Concept and Measurement.* Department of Park, Recreation and Tourism Resources, Michigan State University.

Kauffman, S. (1995) *At Home in the Universe: The Search for the Laws of Self-Organization and Complexity.* New York: Oxford University Press.

Kaye, M. and Taylor, W.G.K. (1997) Expatriate culture shock in China: A study in the Beijing hotel industry. *Journal of Managerial Psychology* 12, 496–510.

Keller, K.L. (1993a) Conceptualizing, measuring, and managing customer-based brand equity. *Journal of Marketing* 57 (1), 1–22.

Keller, K.L. (1993b) Memory retrieval factors and advertising effectiveness. In A.A. Mitchell (ed.) *Advertising Exposure, Memory, and Choice: Advertising and Consumer Psychology* (pp. 11–48) Hillsdale, NJ: Lawrence Erlbaum Associates.

Keller, K.L. (1998) *Strategic Brand Management: Building, Measuring and Managing Brand Equity.* Upper Saddle River, NJ: Prentice Hall.

Keller, K.L. (2003) *Strategic Brand Management: Building, Measuring and Managing Brand Equity* (2nd edn). Upper Saddle River, NJ: Prentice Hall.

Keogh, B. (1990) Public participation in community tourism planning. *Annals of Tourism Research* 17 (3), 449–65.

Keynes, J.M. (1936) *A General Theory of Employment, Interest and Money*. London: Macmillian.

Kilduff, M. and Tsai, W. (2003) *Social Networks and Organizations*. London: Sage Publications.

King, B. and Choi, H.J. (1999) Travel industry structure in fast growing but immature outbound markets: The case of Korea to Australia travel. *International Journal of Tourism Research* 1, 111–22.

King, R. (1990) Policy and process in the modern state. In J. Simmie and R. King (eds) *The State in Action: Public Policy and Politics* (pp. 3–21). London: Pinter

Kleijnen, J.P.C. (1995) Verification and validation of simulation models. *European Journal of Operational Research* 82, 145–62.

Klein, E.E., and Herskovitz, P.J. (2005) Philosophical foundations of computer simulation validation. *Simulation and Gaming* 36 (3), 303–29.

Kleindorfer, G.B., O'Neill, L. and Ganeshan, R. (1998) Validation in simulation: Various positions in the philosophy of science. *Management Science* 44 (8), 1087–99.

Klijn, E. (1996) Analyzing and managing policy processes in complex networks: A theoretical examination of the concept of policy network and its problems. *Administration and Society* 28 (1), 90–119.

Klijn, E.H. and Koppenjan, J.F.M. (2000) Public management and policy networks: Foundations of a network approach to governance. *Public Management* 2 (2), 135–58.

Knoke, D. and Kuklinski, J.H. (1991) Network analysis: Basic concepts. In G. Thompson, J. Frances, R. Levacic and J. Mitchell (eds) *Markets, Hierarchies and Networks* (pp. 173–82). London: Sage Publications.

Knoke, D. (1980) *Political Networks: The Structural Perspective*. Cambridge: Cambridge University Press.

Knoke, D. (1990) *Political Networks: The Structural Perspective*. Cambridge: Cambridge University Press.

Knoke, D. (1993) Networks of elite structure and decision making. *Sociological Methods and Research* 22 (1), 23–45.

Kogut, B. (2000) The network as knowledge: Generative rules and the emergence of structure. *Strategic Management Journal* 21 (3), 405–25.

Konecnik, M. (2006) Croatian-based brand equity for Slovenia as a tourism destination. *Economic and Business Review for Central and South-Eastern Europe* 8 (1), 83–108.

Konecnik, M. and Gartner, W.C. (2007) Customer-based brand equity for a destination. *Annals of Tourism Research* 34 (2), 400–21.

König, D. (1936) *Theorie der endlichen und unendlichen Graphen*. Leipzig: Akademische Verlagsgesellschaft m.b.h.

König, T. (1998) Modelling policy networks. *Journal of Theoretical Politics* 10, 387–8.

Kooiman, J. (1993) Socio-political governance. In J. Kooiman (ed.) *Modern Governance: New Government–Society Interactions* (pp. 1–9). London: Sage.

Kossinets, G. (2006) Effects of missing data in social networks. *Social Networks* 28 (3), 247–68.

Kotler, P., Haider, D.H. and Rein, I. (1993) *Marketing Places: Attracting Investments, Industry, and Tourism to Cities, States and Nations*. New York: Free Press.

Krackhardt, D., Blythe, J. and McGrath, C. (1994) KrackPlot 3.0: An improved network drawing program. *Connections* 17 (2), 53–5.

Krackhardt, D. (1992) The strength of strong ties: The importance of philos in organisations. In N. Nohria and R. G. Eccles (eds) *Networks and Organizations: Structure, Form and Action*. Boston: Harvard Business School Press.

Kreiner, K. and Schulz, M. (1993) Informal collaboration in R and D: The formation of networks across organizations. *Organization Studies* 14, 189–209.

Küppers, G. and Lenhard, J. (2005) Validation of simulation: Patterns in the social and natural sciences. *Journal of Artificial Societies and Social Simulation* 8 (4), Retrieved March 2006 from http://jasss.soc.surrey.ac.uk/8/4/3.html.

Ladd, D.A. and Ward, M.A. (2002) Of environmental factors influencing knowledge transfer. *Journal of Knowledge Management Practice* 3, (http:www.tlainc.com/jkmp3.htm accessed 15/3/04).

Ladeur, K. (2004) Globalization and public governance – A contradiction. In K. Ladeur (ed.) *Public Governance in the Age of Globalisation* (pp. 1–22). Aldershot: Ashgate.

Lambert, J. (1996) *Services and Tourism. Business Networks Business Growth*. Sydney, AusIndustry: Business Networks Program.

Lang, S. (1970) *Linear Algebra*. Reading, MA.: Addison-Wesley Pub. Co.

Langley, R. (1971) *Practical Statistics*. New York: Dover.

Lasswell, H.D. (1936) *Politics; Who Gets What, When, How*. New York, London: Whittlesey House McGraw-Hill.

Latora, V. and Marchiori, M. (2001) Efficient behavior of small-world networks. *Physical Review Letters* 87, 198701.

Laumann, E., Galaskiewicz, J. and Marsden, D. (1978) Community structure as interorganizational linkages. *Annual Review of Sociology* 4, 455–84.

Law, A.M. and Kelton, W.D. (2000) *Simulation Modelling and Analysis* (3rd edn). New York: McGraw-Hill.

Lawrence, M. (2005) Unraveling the complexities of tourism destination systems: Policy networks and issue cycles 1988–2005. Unpublished PhD Thesis. School of Tourism and Hospitality Management, Southern Cross University, Australia.

Lawrence, S. (2000) Context in Web search. *IEEE Data Engineering Bulletin* 23 (3), 25–32.

Laws, E. (1995) *Tourist Destination Management: Issues, Analysis, and Policies*. London: Routledge.

Laws, E. (2003). Towards an analysis of complex tourism systems. A paper presented at the CAUTHE Conference, Coffs Harbour.

Laws, E., Scott, N. and Parfitt, N. (2002) Synergies in destination image management: A case study and conceptualisation. *The International Journal of Tourism Research* 4 (1), 39–55.

Lechner, C. and Dowling, M. (1999) The evolution of industrial districts and regional networks: The case of the biotechnology region Munich/Martinried. *Journal of Management and Governance* 3 (4), 309–38

Lee, S.H., Kim, P.-J. and Jeong, H. (2006) Statistical properties of sampled networks. *Physical Review E* 73, 016102.

Leeuwis, C. (1991) Equivocations on knowledge system theory: An actor-oriented critique. In D. Kuiper and N. Rolling (eds) *The Edited Proceedings of the European Seminar on Knowledge Management and Information Technology* (pp. 107–16) Wageningen: Wagegingen Agricultural University.

Leiper, N. (1989). Tourism and Tourism Systems (Vol. Occasional Paper No 1). Department of Management Systems, Palmerston North Massey University.

Leiper, N. (1990) Partial industrialisation of tourism systems. *Annals of Tourism Research* 17 (4), 600–5.

Leiper, N. (2000) An emerging discipline. *Annals of Tourism Research* 27 (3), 805–9.

Leung, T.K.P., Wong, Y.H. and Wong, S. (1996) A study of Hong Kong businessmen's perceptions of the role 'guanxi' in the People's Republic of China. *Journal of Business Ethics* 15, 749–58.

Levin, S.A. (2003) Complex adaptive systems: Exploring the known, the unknown and the unknowable. *Bulletin of the American Mathematical Society* 40 (1), 3–19.

Levine, J. and Moreland, R. (1990). Progress in small group research. *Annual Review of Psychology*, 41 (1), 585–634.

Levine, S. and White, P.E. (1961) Exchange as a conceptual framework for the study of interorganisational relationships. *Administrative Science Quarterly* 5, 583–601.

Lewin, R. (1999) *Complexity, Life on the Edge of Chaos* (2nd edn). Chicago: The University of Chicago Press.

Li, F. (1995) Corporate networks and the spatial and functional reorganisations of large firms. *Environment and Planning* 27, 1627–45.

Linder S.H. and Rosenau P.V. (2000) Mapping the terrain of the public–private policy partnership. In P.V. Rosenau (ed.) *Public–Private Policy Partnerships* (pp. 1–18). Cambridge MA: MIT Press.

Lindstrand, A. (2003) How to use network experience in ongoing international business. In A. Blomstermo and D.D. Sharma (eds) *Learning in the Internationalisation Process of Firms* (pp. 77–104). Cheltenham: Edward Elgar.

Link, A. and Rees, J. (1991) Firm size, university-based research and the returns to RandD. In Z.J. Acs and D.B. Audretsch (eds) *Innovation and Technological Change: An International Comparison* (pp. 60–70). Hemel Hemstead: Harvester Wheatsheaf.

Lionberger, H.F. and Gwin, P.H. (1991) *Technology Transfer From Researchers to Users*. Missouri: University of Missouri.

Liyanage, S. (1995) Breeding innovation clusters through collaborative research networks. *Technovation* 15 (9), 553–67.

Locke, K. and Golden-Biddle, K. (1997) Constructing opportunities for contribution: Structuring intertextual coherence and 'problematizing' in organizational studies. *Academy of Management Journal* 40 (5), 1023–62.

López-Pintado, D. (2004) *Diffusion in Complex Social Networks* (Working Papers No. AD 2004–33): Instituto Valenciano de Investigaciones Económicas, S.A. (Ivie) Retrieved December, 2005, from http://www.ivie.es/downloads/docs/wpasad/wpasad–2004–33.pdf.

Lorenzoni, G. and Lipparini, A. (1999) The leverage of interfirm relationships as a distinctive organization capability: A longitudinal study. *Strategic Management Journal* 20, 317–38.

Lovelock, B. (2001) Interorganisational relations in the Protected Area–Tourism Policy domain: The influence of macro-economic policy. *Current Issues in Tourism* 4 (2/4), 253–74.

Lynch, P (2000) Networking in the homestay sector. *The Service Industries Journal* 20 (3), 95–116.

Macy, M. and Skvoretz, J. (1998) The evolution of trust and cooperation between strangers: A computational model. *American Sociological Review* 63 (5), 638–60.

Madhavan, R., Koka, B. and Prescott, J. (1998) Networks in transition: How industry events (re) shape interfirm relationships. *Strategic Management Journal* 19 (5), 439–59.

Madsen, T.K., and Servais, P. (1997) The internationalisation of born globals: An evolutionary process? *International Business Review* 6 (6), 561–83.

Magnusson, J. and Nilsson, A. (2003) To facilitate or intervene – A study of knowledge management practice in SME networks. *Journal of Knowledge Management Practice* 4 (http:www.tlainc.com/jkmp4.htm accessed 15/3/04).

Maier, F.H., and Größler, A. (2000) What are we talking about? – A taxonomy of computer simulations to support learning. *System Dynamics Review*,16 (2), 135–48.

Malhotra, N.K., Agarwal, J. and Peterson, M. (1996) Methodological issues in cross-cultural marketing research: A state-of-the-art review. *International Marketing Review* 13, 7–43.

Mandell, M.P. (1999) The impact of collaborative efforts: Changing the face of public policy through networks and network structures. *Policy Studies Review* 16 (1).

Manning, E. (1998) *Governance for Tourism: Coping with Tourism in Impacted Destinations*. Ottawa: Centre for a Sustainable Future.

Mantegna, R.N., and Stanley, H.E. (2000) *Introduction to Econophysics: Correlations and Complexity in Finance*. Cambridge: Cambridge University Press.

March, J.G. and Olsen, J.P. (1995) *Democratic Governance*. New York: Free Press.

March, R. (1997) An exploratory study of buyer–supplier relationships in international tourism: The case of Japanese wholesalers and Australian suppliers. *Journal of Travel and Tourism Marketing* 6, 55–68.

Marcouiller, D.W. (1997) Toward integrative tourism planning in rural America. *Journal Of Planning Literature* 11 (3), 337–57.

Margerum, R.D. (2002) Collaborative planning: Building consensus and building a distinct model for practice. *Journal of Planning Education and Research* 21 (3), 237–53.

Marsden, P.V. (1990) Network Data and Measurement. *Annual Review of Sociology* 16, 435–63.

Marsh, D. (1998) *Comparing Policy Networks*. Buckingham: Open University Press.

Marsh, I. (2002) Governance in Australia: Emerging issues and choices. *Australian Journal of Public Administration*, 61 (2), 3–9.

Martin C. (2002) *Technology Diffusion within Small and Medium Enterprises in Australia: Report on the Effectiveness of Dissemination Methods*. Canberra: Department of Industry Tourism and Resources.

MATLAB. (2004) MATLAB (Version 7, R14–2004) Natick, MA: The Matworks Inc.

Matteo, T.D., Aste, T. and Gallegati, M. (2005) Innovation flow through social networks: Productivity distribution in France and Italy. *The European Physical Journal B – Condensed Matter and Complex Systems* 47 (3), 459–66.

Mayntz, R. (1993) Policy–Netzwerke und die Logik von Verhandlungssystemen. In A. Héritier (Hrsg.) *Policy Analyse. Kritik und Neuorientierung. Politische Vierteljahresschrift, Sonderheft* 24 (pp. 39–56) Opladen: Westdeutscher Verlag.

Mayntz, R. (1999) *New Challenges to Governance Theory*. Mannheim: Max-Plank Institut für Gesellschaftsforschung.

McClure, M. (2004) *The Wonder Country: Making New Zealand Tourism*. Auckland: Auckland University Press.

McKercher, B. (1998) The effects of market access on destination choice. *Journal of Travel Research* (August), 39–47.

McKercher, B. (1999) A chaos approach to tourism. *Tourism Management* 20, 425–34.

McLaren, T., Head, M., and Yuan, Y. (2002) Supply chain collaboration alternatives: understanding the expected costs and benefits. *Internet Research: Electronic Networking Applications and Policy* 12 (4), 348–64.

McPherson, M., Smith-Lovin, L. and Cook, J. (2001) Birds of a feather: Homophily in social networks. *Annual Review of Sociology* 27, 415–44.

Medeiros de Araujo, L. and Bramwell, B. (1999) Stakeholder assessment and collaborative tourism planning: The case of Brazil's Costa Dourada Project. *Journal of Sustainable Tourism* 7, 356–365.

Metropolis, N., Rosenbluth, A.W., Rosenbluth, M.N., Teller, A.H. and Teller, E. (1953) Equation of state calculation by fast computing machines. *Journal of Chemical Physics* 21, 1087–91.

Micela, AL. Roberti, P. and Jucucci, G. (2002) From individual tourism organization to a single virtual tourism organization for destination management. In K.W. Wober, A.J. Frew and M. Hitz (eds) *Information and Communication Technologies in Tourism* (pp. 87–96). Vienna: Springer Verlag.

Middleton, V.T.C. (1989) Tourist product. In L. Moutinho and S.F. Witt (eds) *Tourism Marketing and Management Handbook* (pp. 573–6). Englewood Cliffs, NJ: Prentice Hall.

Middleton, V.T.C., and Clarke, J. (2001) *Marketing in Travel and Tourism* (3rd edn). Oxford: Butterworth-Heinemann.

Miles, M. and Huberman, A. (1994) *Qualitative Data Analysis: A Sourcebook of New Methods*. London: Sage Publications.

Milgram, S. (1967) The small world problem. *Psychology Today* 1, 60–7.

Mill, R. and Morrison, A. (2002). *The Tourism System*. Dubuque, Iowa: Kendall Hunt.

Miller, W.B. (1958) Inter-institutional conflict as a major impediment to delinquency prevention. *Human Organisation* 17, 20–3.

Mirowski, P. (1989) *More Heat than Light: Economics as Social Physics, Physics as Nature's Economics*. Cambridge: Cambridge University Press.

Mitchell, J.C. (1969) The concept and use of social networks. In J. C. Mitchell (ed.) *Social Networks in Urban Situations* (pp. 1–50) Manchester: University of Manchester Press.

Mizruchi, M.S. (1994) Social network analysis: Recent achievements and current controversies. *Acta Sociologica* 37, 329–43.

Möllering, G. (2001) The nature of trust: From Georg Simmel to a theory of expectation, interpretation and suspension. *Sociology* 35 (02), 403–20.

Money, R.B. and Crotts, J.C. (2000) Buyer behaviour in the Japanese travel trade: Advancements in theoretical frameworks. *Journal of Travel and Tourism Marketing* 9, 1–19.

Monge, P. and Contractor, F.J. (1999) Emergence of communication networks. In F. M. Jablin and L.L. Putnam (eds) *Handbook of Organizational Communication* (2nd edn) (pp. 440–502). Thousand Oaks, CA: Sage.

Monge, P. and Contractor, F.J. (2003) *Theory of Communication Networks*. New York: Oxford University Press.

Monge, P.R. (1987) The network level of analysis. In C.R. Berger and S.H. Chaffee (eds) *Handbook of Communication* Science (239–70). Newbury Park, CA: Sage.

Moreno, J.L. (1934) *Who Shall Survive?* Washington, DC: Nervous and Mental Disorders Publishing Co.

Morgan, G. (1988) *Riding the Waves of Change: Developing Managerial Competencies for a Turbulent World*. San Francisco: Jossey-Bass.

Morgan, N. and Pritchard, A. (1998) *Tourism Promotion and Power: Creating Images, Creating Identities*. Chichester, New York: Wiley.

Morgan, N. and Pritchard, A. (1999) *Power and Politics at the Seaside: The Development of Devon's Resorts in the Twentieth Century*. University of Exeter Press.

Morgan, N., Pritchard, A. and Piggott, R. (2002) New Zealand, 100% pure. The creation of a powerful niche destination brand. *Journal of Brand Management* 9 (4/5), 335–54.

Morgan, N., Pritchard, A. and Piggott, R. (2003) Destination branding and the role of stakeholders: The case of New Zealand. *Journal of Vacation Marketing* 9 (3), 285–99.

Morrison, A.M. and Anderson, D.J. (2002) Destination branding. Paper presented at the Missouri Association of Convention and Visitor Bureaus Annual Meeting.

Morrison, A., Lynch, P. and Johns, N. (2004) International tourism networks. *International Journal of Contemporary Hospitality Management* 16 (3), 197–202.

Mossa, S., Barthélémy, M., Stanley, H.E. and Amaral, L.A.N. (2002) Truncation of power law behavior in 'scale-free' network models due to information filtering. *Physical Review Letters* 88 (13), 138701.

Murphy, P.E. (1985) *Tourism: A Community Approach*. New York: Methuen.

Mytelka, L.K. (2002) Clustering, long distance partnerships and the SME: A study of the French biotechnology sector. Paper presented at the Fourth Annual Meeting of the Innovation Systems Research Network (ISRN) Quebec, 9–10 May.

Naylor, T.H. and Finger, J.M. (1967) Verification of computer simulation models. *Management Science* 14 (2), B92–B101.

Newman, M.E.J. (2002) The structure and function of networks. *Computer Physics Communications* 147, 40–5.

Newman, M.E.J. (2003) The structure and function of complex networks. *SIAM Review* 45 (2), 167–256.

Newman, M.E.J. (2004) Analysis of weighted networks. *Physical Review E* 70, 056131.

Newman, M.E.J. and Girvan, M. (2004) Finding and evaluating community structure in networks. *Physical Review E* 69, 26113.

Nölke, A. (2004) Limits to democratic network governance: The case of transnational politics (Paper prepared for the Conference on Democratic Network Governance. Copenhagen, October 21–2).

Nonaka, I. (1991) The knowledge creating company. *Harvard Business Review* 69 (6), 96–104.

Nordin, S. (2003) *Tourism Clustering and Innovation: Paths to Economic Growth and Development*. Oestersund: European Tourism Research Institute, Mid-Sweden University.

Novelli, M., Schmitz, B. and Spencer, T. (2006) Networks, clusters and innovation in tourism: A UK experience. *Tourism Management* 27 (6), 1141–52.

O'Connor, P., and Murphy, J. (2004) Research on information technology in the hospitality industry. *Hospitality Management* 23, 473–84.

Observatory of European SMEs (2003) *SMEs in Europe 2003 [online]*. Observatory Report N° 7. European Commission, Available from: http://ec.europa.eu/ [Accessed 23 January 2005].

Odlyzko, A.M. (2000) The history of communications and its implications for the Internet. Retrieved September, 2006, from http://ssrn.com/abstract=235284.

Ohmae, K. (1995) *The End of the Nation State. The Rise of Regional Economies. How New Engines of Prosperity are Reshaping Global Markets*. London: HarperCollins Publishers.

Oliver, C. (1988) The collective strategy framework: An application to competing predictions of isomorphism. *Administrative Science Quarterly* 33, 543–61.

Olson, M. (1965) *The Logic of Collective Action: Public Goods and the Theory of Groups.* Cambridge, MA: Harvard University Press.

O'Reilly, P. (1988) Methodological issues in social support and social network research. *Social Science and Medicine* 26 (8), 863–73.

Organisation for Economic Cooperation and Development (2001) *Cities and Regions in the New Learning Economy.* Paris: OECD.

Osborne, D. and Gaebler, T. (1992) *Reinventing Government. How the Entrepreneurial Spirit is Transforming the Public Sector.* New York: Penguin.

Ostrom, E. (1990) *Governing the Commons: The Evolution of Institutions for Collective Action.* Cambridge: Cambridge University Press.

Palay, T. (1984) Comparative institutional economics: The governance of rail freight contracting. *Journal of Legal Studies* 13, 265–88.

Palmer, A. (1998) Evaluating the governance style of marketing groups. *Annals of Tourism Research* 25 (1), 185–201.

Palmer, A. and Bejou, D. (1995) Tourism destination marketing alliances. *Annals of Tourism Research* 22 (3), 616–29.

Palmer, A. (1996) Linking external and internal relationship building in networks of public and private sector organizations: A case study. *International Journal of Public Sector Management*, 9 (3), 51–60.

Pan, G.W., Sparks B.A. and Fulop, L. (2007) Cross-cultural partner relationships in travel trade: A Sino–Australian case study. *Journal of Travel Research* 46 (2), 195–206.

Pappi, F.U. (1993) Policy–Netze: Erscheinungsform moderner Politiksteuerung oder methodischer Ansatz? In A. Héritier (Hrsg.) *Policy Analyse. Kritik und Neuorientierung.* Politische Vierteljahresschrift, Sonderheft 24 (pp. 84–94) Opladen: Westdeutscher Verlag.

Park, H.W. (2003) Hyperlink network analysis: A new method for the study of social structure on the Web. *Connections* 25 (1), 49–61.

Park, H.W. and Thelwall, M. (2003) Hyperlink analyses of the World Wide Web: A review. *Journal of Computer Mediated Communication [On–line]* 8 (4), Retrieved March 2006 from http://jcmc.indiana.edu/vol8/issue4/park.html.

Parkhe, A., Wasserman, S. and Ralston, D. (2006) New frontiers in network theory development. *Academy of Management Review* 31 (3), 560–8.

Pastor-Satorras, R. and Vespignani, A. (2001) Epidemic spreading in scalefree networks. *Physical Review Letters* 86 (14), 3200–3.

Pastor-Satorras, R. and Vespignani, A. (2004) *Evolution and Structure of the Internet: A Statistical Physics Approach.* Cambridge: Cambridge University Press.

Pavard, B. and Dugdale, J. (2000) The contribution of complexity theory to the study of socio-technical cooperative systems. Third International Conference on Complex Systems, Nashua, NH, May 21–26. Retrieved October, 2005, from http://www–svcict.fr/cotcos/pjs/.

Pavlovich, K. (2001) The twin landscapes of Waitomo: Tourism network and sustainability through the Landcare Group. *Journal of Sustainable Tourism* 9 (6), 491–504.

Pavlovich, K. (2003) The evolution and transformation of a tourism destination network: The Waitomo Caves, New Zealand. *Tourism Management* 2(2), 203–16.

Pavlovich, K. and Kearins, K. (2004) Structural embeddedness and community-building through collaborative network relationships. *M@n@gement* 7 (3), 195–214.

Pearce, D. (1996) Tourist organizations in Sweden. *Tourism Management* 17 (6), 413–24.

Pearce, P.L., Kim, E. and Lussa, S. (1998) Facilitating tourist–host social interaction: An overview and assessment of the culture assimilator. In E. Laws, B. Faulkner and G. Moscardo (eds) *Embracing and Managing Change in Tourism: International Case Studies*. London and New York: Routledge.

Pechlaner, H. Abfalter, D. and Raich, F. (2002) Cross-border destination management systems in the Alpine region: The role of knowledge networks on the example of AlpNet. In R.B. Bouncken and S. Pyo (eds) *Knowledge Management in Hospitality and Tourism* (pp. 89–108). New York: Haworth.

Pechlaner, H., Tallinucci, V., Abfalter, D. and Rienzner, H. (2003) Networking for small island destinations: The case of Elba. In A.J. Frew, M. Hitz and P. O'Connor (eds) *Information and Communication Technologies in Tourism* (pp. 105–14). Wien: Springer.

Pennock, D.M., Flake, G.W., Lawrence, S., Glover, E.J. and Giles, C.L. (2002) Winners don't take all: Characterizing the competition for links on the web. *Proceedings of the National Academy of the Sciences of the USA* 99 (8), 5207–11.

Pentland, B. (1999) Building process theory with narrative: From description to explanation. *Academy of Management Review* 24, 711–24.

Perryman, A.A. and Combs, J.G. (2005) Collaborative entrepreneurship: How networked firms use continuous innovation to create economic wealth. By Raymond E. Miles, Grant Miles, and Charles Snow (book review). *The International Entrepreneurship and Management Journal* 1 (3), 399–401.

Pesämaa, O. and Skurla, J.A. (2003) Secret ties as a way to succeed: Describing and exploring relations among successful tourism managers. Paper presented at the Perspectives on Tourism in Nordic and Other Peripheral Areas International Conference, Umeå, Sweden (21–24 August 2003)

Peters, M., Weiermair, K. and Withalm, J. (2002) Small and Medium Sized Enterprises Alliance Through Research in Tourism (SMART–UP). In K.W. Wober, A.J Frew and M. Hitz (eds) *Information and Communication Technologies in Tourism* (pp. 145–56). Vienna: Springer Verlag.

Pforr, C. and Thompson, G. (2004) *The Structure of the Australian Tourism Industry: The Case of Western Australia*.

Pforr, C. (2001) Tourism policy in Australia's Northern Territory: A policy process analysis of its tourism development masterplan. *Current Issues in Tourism* 4 (2), 275–307.

Pforr, C. (2002) The 'makers and shapers' of tourism policy in the Northern Territory of Australia: A policy network analysis of actors and their relational constellations. *Journal of Hospitality and Tourism Research* 9 (2), 134–51.

Pforr, C. (2004) Sustainable tourism – Governance through policy networks. In R. MacLellan, T. Baum, A. Goldsmith, J. Kokkranikal, E. Losekoot, S. Miller, A. Morrison, D. Nickson, J.S. Taylor and K. Thompson (eds) *Tourism: State of the Art II* (Conference Proceedings, 27–30 June) (pp. 1–19). Glasgow: The Scottish Hotel School, University of Strathclyde.

Pforr, C. (2005) Three lenses of analysis for the study of tourism public policy. *Current Issues in Tourism* 8 (4): 323–43.

Pforr, C. (2006a) Tourism policy in the making: An Australian network study. *Annals of Tourism Research* 33 (1), 87–108.

Pforr, C. (2006b) Regional tourism in transition: Western Australia's new concept for regional tourism. Paper presented at the International Conference of Trends,

Impacts and Policies on Tourism Development, Amoudara, Heraklion, Crete, June 15–18.

Pforr, C. and Megerle, A. (2005) Geotourism: A perspective from South-West Germany. In R. Dowling and D. Newsome (eds) *Geotourism: Sustainability – Impacts and Opportunities* (pp. 118–139). Oxford: Butterworth-Heinemann.

Phillips, N., Lawrence, T.B. and Hardy, C. (2000) Inter-organizational collaboration and the dynamics of institutional fields. *Journal of Management Studies* 37 (1), 23–43.

Pierre, J. (ed.) (2000) *Debating Governance: Authority, Steering and Democracy*. Oxford: Oxford University Press.

Pike, S. (2004) *Destination Marketing Organisations*. Amsterdam: Elsevier.

Podolny, J.M. and Page, K. (1998) Network forms of organization. *Annual Review of Sociology* 24, 57–76.

Poincaré, H. (1883) Sur certaines solutions particulieres du probleme des trois corps. *Comptes-Rendus de l'Académie des Sciences Paris* 97, 251–2.

Poincaré, H. (1884) Sur certaines solutions particulieres du probleme des trois corps. *Bulletin Astronomique* 1, 63–74.

Polanyi, M (1966) *The Tacit Dimension*. New York: Doubleday.

Poon, A. (1993) *Tourism, Technology and Competitive Strategies*. Oxford: CAB International.

Porter, M. (1990) *The Competitive Advantage of Nations*. New York: Basic Books.

Porter, M. (1998) Clusters and the new economics of competition. *Harvard Business Review*, 76 (6), 77–90.

Powell, W. Koput, K. and Smith-Doerr, L. (1996) Interorganizational collaboration and the locus of innovation: Networks of learning in biotechnology. *Administrative Science Quarterly* 41, 116–45.

Powell, W., White, D., Koput, K. and Owen-Smith, J. (2005) Network dynamics and field evolution: The growth of interorganisational collaboration in the life sciences. *American Journal of Sociology* 110 (4), 1132–1205.

Prensky, M. (2001) *Digital Game-Based Learning*. New York: McGraw–Hill.

Pressman, J.L. and Wildavsky, A. (1984) *Implementation: How Great Expectations in Washington are Dashed in Oakland; or, Why it's Amazing that Federal Programs Work at All, this being a Saga of the Economic Development Administration as told by Two Sympathetic Observers who Seek to Build Morale on a Foundation of Ruined Hopes* (3rd edn). Berkeley: University of California Press.

Prideaux, B. (2004) The resort development spectrum: The case of the Gold Coast, Australia. *Tourism Geographies* 6 (1), 26–58.

Prideaux, B. and Cooper, C. (2002) Marketing and destination growth: A symbiotic relationship or simple coincidence? *Journal of Vacation Marketing* 9 (1), 35–48.

Pyatt T. (1995) Survey and speculations: Business networks and dyads studies: Theory and practice in South East Asia. *Journal of Far East Business* 1, 1–14.

Pye, L. w. (1985) *Asian Power and Politics: the Cultural Dimensions of Authority*. Cambridge, MA: Belknap Press.

Rafiei, D. and Curial, S. (2005) Effectively visualizing large networks through sampling. *Proceedings of the IEEE Visualization Conference (VIS'05)*, Minneapolis, Oct. 2005, 48–56.

Ragowsky, A., Stern., M., and Adams. A.D. (2000) Relating benefits from using IS to an organization's operating characteristics: Interpreting results from two countries. *Journal of Management Information Systems* 16, 4.

Ravasz, E. and Barabási, A.-L. (2003) Hierarchical organization in complex networks. *Physical Review E* 67, 026112.

Redding, S.G. (1993) *The Spirit of Chinese Capitalism*. Berlin, New York: Walter de Gruyter.

Reed, M.G. (1997) Power relations and community-based tourism planning. *Annals of Tourism Research* 24 (3), 566–91.

Reid, W. (1964) Interagency coordination in delinquency prevention and control. *Social Service Review* 38, 418–28.

Reisinger, Y. and Turner, L. (1997) Cross-cultural differences in tourism: Indonesian tourists in Australia. *Tourism Management* 18, 139–47.

Reisinger, Y. and Turner, L. (1998) Cross-cultural differences in tourism: A strategy for tourism marketers. *Journal of Travel and Tourism Marketing* 7, 79–106.

Reynolds, C. (1987) Flocks, herds, and schools: a distributed behavioral model. *Computer Graphics* 21, 25–34.

Rhodes, R.A.W. (1990) Policy networks: A British perspective. *Journal of Theoretical Politics* 2, 293–317.

Rhodes, R.A.W. (1997) *Understanding Governance. Policy Networks, Governance, Reflexivity and Accountability*. Buckingham: Open University Press.

Rhodes, R.A.W. (2002) Putting people back into networks. *Australian Journal of Political Science* 37 (3), 399–416.

Richards, F. and Carson, D. (2006) Foundations of innovation: The role of local government in the production and distribution of knowledge in regional tourism systems. Paper presented at the 2006 CAUTHE Conference, Melbourne

Richardson, K.A. (2003) On the limits of bottom-up computer simulation: Towards a nonlinear modeling culture. *Proceedings of the 36th Hawaii International Conference on System Sciences (HICSS'03)*, Hawaii, USA.

Ring, P.S., and van de Ven, A.H. (1994) Developmental processes of cooperative interorganizational relationships. *The Academy of Management Review* 19 (1), 90–118.

Ritchie, J.R.B. and Ritchie, J.R.R. (1998) The branding of tourism destinations. Paper presented at the Annual Congress of the International Association of Scientific Experts in Tourism, Marrakech, Morocco.

Roberts, B.H. and Enright, M.J. (2004) Industry clusters in Australia: Recent trends and prospects. *European Planning Studies* 12 (1) January, 99–121.

Robson, J. and Robson, I. (1996) From shareholders to stakeholders: Critical issues for tourism marketers. *Tourism Management* 17 (7) November, 533–40.

Rodgers, G.J. and Darby-Dowman, K. (2001) Properties of a growing random directed network. *The European Physical Journal B* 23, 267–71.

Rogers, E. (1995) *Diffusion of Innovations* (4th edn). New York: Free Press.

Rogers, E.M., and Shoemaker, F.F. (1971) *Communication of Innovations: A Cross Cultural Approach*. New York: Free Press.

Rosenblueth, A. and Wiener, N. (1945) The role of models in science. *Philosophy of Science* 7 (4), 316–21.

Rowley, T.J. (1997) Moving beyond dyadic ties: A network theory of stakeholder influences. *Academy of Management Review* 22 (4), 887–910.

Russell, R. (2005) Chaos theory and its application to the Tourism Area Life Cycle Model. In R.W. Butler (ed.) *The Tourism Area Life Cycle, Vol. 2: Conceptual and Theoretical Issues* (pp. 164–80). Clevedon: Channel View.

Russell, R. and Faulkner, B. (1998) Reliving the destination life cycle in Coolangatta. In E. Laws, B. Faulkner and G. Moscardo (eds) *Embracing and*

Managing Change in Tourism: International Case Studies (pp. 95–115). London: Routledge.

Russell, R. and Faulkner, B. (1999) Movers and shakers: Chaos makers in tourism development. *Tourism Management*, 20 (4), 411–23.

Saari, D.G. (1995) Mathematical complexity of simple economics. *Notices of the American Mathematical Society* 42 (2), 222–30.

Sabatier, P. (1987) Knowledge, policy-oriented learning, and policy change. *Knowledge: Creation, Diffusion and Utilization* 8 (4), 649–92.

Said, E.W. (1978) *Orientalism*. London: Routledge and Kegan Paul.

Saramäki, J., Kivelä, M., Onnela, J.-P., Kaski, K. and Kertész, J. (2006) Generalizations of the clustering coefficient to weighted complex networks (Preprint arxiv/cond–mat/0608670): ArXiv e–prints archive. Retrieved October, 2006, from http://arxiv.org/abs/cond–mat/0608670.

Sautter, E.T. and Leisen, B. (1999) Managing stakeholders. A tourism planning model. *Annals of Tourism Research* 26 (2), 312–28.

Saxena, G. (2005) Relationships, networks and the learning regions: case evidence from the Peak District National Park. *Tourism Management* 26 (2), 277–89.

Saxena, G. (2006) Beyond mistrust and competition – the role of social and personal bonding processes in sustaining livelihoods of rural tourism businesses: A case of the Peak District National Park. *International Journal of Tourism Research* 8, 263–77.

Scarini, P (1996) Elaboration of the Swiss Agricultural Policy for the Gatt Negotiations: A network analysis. *Swiss Journal of Sociology* 22 (1), 85–115.

Schmid, A. (2005) What is the truth of simulation? *Journal of Artificial Societies and Social Simulation* 8 (4) Retrieved June, 2006, from http://jasss.soc.surrey.ac.uk/8/4/5.html.

Schneider, V. (1992) The structure of policy networks. A comparison of the 'chemicals control' and 'telecommunications'. *European Journal of Political Research*, 21, 109–29.

Schneider, V. (2005) Policy networks and the governance of complex societies. Paper presented at the conference on 'Network Societies and Postindustrial Identities' at the University of Tübingen, Germany, Post-Graduate Research Programme 'Global Challenges – Transnational and Transcultural Approaches', Tübingen, October 28–30.

Schubert, K. (1991) *Politikfeldanalyse*. Opladen: Leske and Budrich.

Schumann, W. (1993) Die EG als neuer Anwendungsbereich für die Policy–Analyse: Möglichkeiten und Perspektiven der konzeptionellen Weiterentwicklung. In A. Héritier (ed.) *Policy Analyse. Kritik und Neuorientierung*. Politische Vierteljahresschrift, Sonderheft 24 (394–431) Opladen: Westdeutscher Verlag.

Schumpeter, J.A. (1950) *Capitalism, Socialism and Democracy* (3rd edn). New York: Harper and Row.

Scott, J. (1996) Software review: A toolkit for social network analysis. *Acta Sociologica* 39, 211–16.

Scott, J. (2000) *Social Network Analysis: A Handbook* (2nd edn). London: Sage Publications.

Scott, J. (ed.) (2002) *Social Networks: Critical Concepts in Sociology*. London: Routledge.

Scott, N. and Cooper, C. (2006) Network analysis as a research tool for understanding tourism destinations. Paper presented at the Cutting Edge Research in Tourism conference, University of Surrey 6th –9th June.

Seary, A.J., and Richards, W.D. (2003) Spectral methods for analyzing and visualizing networks: an introduction. In R. Breiger, K. Carley and P. Pattison (eds) *Dynamic Social Network Modeling and Analysis* (pp. 209–28). Washington, DC: National Academy Press.

Selin, S. (1993) Collaborative alliances: New interorganisational forms in tourism. *Journal of Travel and Tourism Marketing* 2 (2/3), 217–27.

Selin, S. and Beason, K. (1991) Interorganisational relations in tourism. *Annals of Tourism Research* 18 (4), 639–52.

Selmer, J. (1997) *Cross-Cultural Management in China: Current Issues and Emerging Trends*. Hong Kong: Hong Kong Baptist University.

Shah, P.P. (2000) Network destruction: The structural implications of downsizing. *Academy of Management Journal* 43 (1), 101–12.

Shalizi, C.R. (2006) Methods and techniques of complex systems science: An overview. In T.S. Deisboeck and J.Y. Kresh (eds) *Complex Systems Science in Biomedicine* (pp. 33–114). New York: Springer.

Shao, J. (1999) *Mathematical Statistics*. Heidelberg: Springer.

Shapira, P. and Rosenfield, L. (1996) An overview of technology diffusion policies and programs to enhance the technological absorptive capabilities of Small and Medium Enterprises. Retrieved accessed 4/06/03, from www.prism.gatech.edu/~jy5/pubs/oecdtech.htm

Sharma, D. and Johnson, J. (1987) Technical consultancy in internationalisation. *International Marketing Review* 4, 20–9.

Shaw, G. and Williams, A.M. (2002) *Critical Issues in Tourism: A Geographical Perspective* (2nd edn). Oxford: Blackwell Publishers.

Sheskin, D.J. (2000) *Handbook of Parametric and Nonparametric Statistical Procedures*. Boca Raton, FL: ChapmanandHall/CRC.

Shih, H.-Y. (2005) Network characteristics of drive tourism destinations: An application of network analysis in tourism. *Tourism Management* 26 (2), 277–89.

Shih, H.-Y. (2006) Network characteristics of drive tourism destinations: An application of network analysis in tourism. *Tourism Management* 27 (5), 1029–39.

Shin, Y. (2006) Collaboration and power relations among stakeholders in a community festival: The case of the Andong Mask Dance festival, South Korea. University of Waterloo Dept of Geography: UMI Dissertation Services.

Shulman, N. (1976) Network analysis: a new addition to an old bag of tricks. *Acta Sociologica* 19 (4), 307–23.

Shuman, J., Twombly, J. and Rottenberg, D. (2001) *Collaborative Communities: Partnering for Profit in the Networked Economy*. Chicago: Dearborn Trade.

Simmons, D.G. (1994) Community participation in tourism planning. *Tourism Management* 15 (Issue 2), 98.

Simon, H.A. (1955) On a class of skew distribution functions. *Biometrika* 42, 425–40.

Simon, H.A. (1997) *Models of Bounded Rationality: Empirically Grounded Economic Reason*. Cambridge, MA: MIT Press.

Siu, R.G.H. (1985) *The Craft of Power*. Malabar, FL: R.E. Krieger Pub. Co.

Skopal, T., Snášel, V., Svátek, V. and Krátký, M. (2003) Searching the Internet using topological analysis of Web pages. *Proceedings of the 2003 International Conference on Communications in Computing (CIC'03)*, Las Vegas, NV, 271–7.

Smeral, E. (1998) The impact of globalization on small and medium enterprises: New challenges for tourism policies in European countries. *Tourism Management* 19 (4), 371–80.

Smith, P.G. and Reinertsen, D.G. (1991) *Developing Products in Half the Time*. New York: Van Rostrand Reinhold.

Smith, V.A., Jarvis, E.D. and Hartemink, A.J. (2003) Influence of network topology and data collection on network inference. *Proceedings of the Pacific Symposium on Biocomputing 2003*, Kauai, Hawaii, 164–75.

Smith, V.L. (1989) *Hosts and Guests: The Anthropology of Tourism* (2nd edn). Philadelphia: University of Pennsylvania Press.

Smith, V.L., and Brent, M. (2001) *Hosts and Guests Revisited: Tourism Issues of the 21st Century*. New York: Cognizant Communication Corp.

Soliman, I. (2001) Collaboration and the negotiation of power. *Asia-Pacific Journal of Teacher Education* 29 (3), 219–34.

Sørensen, E. and Torfing, J. (2004) *Making Governance Networks Democratic*. Working Paper 2004. 1 Roskilde: Centre for Democratic Network Governance.

Sørensen, E. and Torfing, J. (2005) The democratic anchorage of governance networks. *Scandinavian Political Studies* 28 (3), 195–218.

Speak, K.D. (1996) The challenge of health care branding. *Journal of Health Care Marketing* 16 (4), 40–2.

Stacey, R.D. (1996) *Complexity and Creativity in Organizations*. San Francisco: Berrett-Koehler.

Stanislaw, H. (1986) Tests of computer simulation validity: what do they measure? *Simulation and Games* 17 (2), 173–91.

Stauffer, D. (2003) Sociophysics simulations. *Computing in Science and Engineering* 5 (3), 71–75.

Stauffer, D. (2004) Introduction to statistical physics outside physics. *Physica A* 336 (1–2), 1–5.

Steinfield, C., Bouwman, H. and Adelaar, T. (2001) Combining physical and virtual channels: Opportunities, imperatives and challenges. 14th Bled Electronic Commerce Conference, Bled, Slovenia (June 25–26). Retrieved December, 2005, from http://www.msu.edu/~adelaar/pubs/bledfinal.pdf.

Sterman, J.D. (2000) *Business Dynamics; Systems Thinking and Modelling for a Complex World*. New York: McGraw-Hill.

Stevens, B. (1988) Co-operative activities in competitive markets. In *Tourism Research: Exploring Boundaries* (pp. 139–41). Montreal: Travel and Tourism Research Association.

Stokes, R. (2006) Network-based strategy making for events tourism. *European Journal of Marketing* 40 (5/6), 682–95.

Stokowski, P.A. (1992) Social networks and tourist behavior. *American Behavioral Scientist* 36 (2), 212–21.

Stumpf, M.P.H., and Wiuf, C. (2005) Sampling properties of random graphs: The degree distribution. *Physical Review E* 72, 036118.

Stumpf, M.P.H., Ingram, P.J., Nouvel, I. and Wiuf, C. (2005a) Statistical model selection methods applied to biological networks. *Transactions in Computational Systems Biology* 3, 65–77.

Stumpf, M.P.H., Wiuf, C. and May, R.M. (2005b) Subnets of scale-free networks are not scale-free: Sampling properties of networks. *Proceedings of the National Academy of the Sciences of the USA* 102 (12), 4221–4.

Suleiman, R., Troitzsch, K.G. and Gilbert, N. (eds) (2000) *Tools and Techniques for Social Science Simulation*. Heidelberg: Physica-Verlag.

Sundbo, J., Orfila-Sintes, F. and Sorensen, F. (in press) The innovative behaviour of tourism firms: Comparative studies of Denmark and Spain. *Research Policy*.

Sweeting, B. (1995) Competition, co-operation and changing the manufacturing infrastructure. *Regional Studies* 29 (1), 87–94.

Swift, J.S. (1999) Cultural closeness as facet of cultural affinity. *International Marketing Review* 16, 182–201.

Szarka, J. (1990) Networking and small firms. *International Small Business Journal* 8 (2), 10–21.

Sznajd-Weron, K. and Sznajd, J. (2000) Opinion evolution in closed community. *International Journal of Modern Physics C* 11 (6), 1157–65.

Sztompka, P. (1999) *Trust: A Sociological Theory*. Cambridge: Cambridge University Press.

Tallinucci, V. and Testa, M. (2006) *Marketing per le isole*. Milano: Franco Angeli.

Teare, R. (1992) Promoting service excellence through service branding. *International Journal of Contemporary Hospitality Management* 4 (1), III.

Tesfatsion, L. and Judd, K.L. (2006) *Handbook of Computational Economics, Volume 2: Agent-Based Computational Economics*. Amsterdam: North-Holland.

Thatcher, M. (1998) The development of policy network analyses from modest origins to overarching frameworks. *Journal of Theoretical Politics* 10 (4), 389–416.

Thibaut, J. and Kelley, H.H. (1959) *The Social Psychology of Groups*. New York: John Wiley.

Thompson, G. and Pforr, C. (2005) Policy networks and good governance – A discussion. *Working Paper Series 2005: 1*. Perth: School of Management, Curtin University of Technology.

Thrift, N. (1996) *Spatial Formations*. London: Sage Publications.

Tichy, N.M., Tushman, M.L. and Fombrum, C. (1979) Social network analysis for organizations. *Academy of Management Review* 4 (4), 507–19.

Timothy, D. (1998) Cooperative tourism planning in a developing destination. *Journal of Sustainable Tourism* 6 (1), 52–68.

Tinsley, R. and Lynch, P. (2001) Small tourism business networks and destination development. *Hospitality Management* 20 (4), 367–78.

Tixier, M. (2000) Communication and management styles in Australia: Understanding the changing nature of its corporate affairs. *Cross Cultural Management* 7, 12–22.

Tjosvold, D. (1986) The dynamics of interdependence in organisations. *Human Relations* 39 (6), 517–40.

Toroczkai, Z. and Eubank, S. (2005) Agent-based modeling as decision-making tool. *The Bridge* 35 (4), 22–27.

Tourism Forecasting Council (2005) *Inbound Tourism Forecasts – 2005 to 2014*. Canberra: Tourism Research Australia.

Tourism NSW (2003) *Hunter Valley Statistics*. Tourism NSW.

Tourism NSW (2006) *Domestic Overnight Travel to NSW*. Tourism NSW.

Tourism Queensland 2004) *Gold Coast Regional Update*. Brisbane: Tourism Queensland.

Tremblay, P. (1998) The economic organization of tourism. *Annals of Tourism Research* 24 (4), 837–59.

Treuren, G. and Lane, D. (2003) The tourism planning process in the context of organised interests, industry structure, state capacity, accumulation and sustainability. *Current Issues in Tourism* 6 (1), 1–22.

Tribe, J. (1997) The indiscipline of tourism. *Annals of Tourism Research* 24 (3), 638–57.

Tsui, A.S., and Fahr, J. (1997) Where guanxi matters: Relational demography and guanxi in the Chinese context. *Work and Occupations* 24 (1), 56–79.

Tsui, A.S., Farh, J.L. and Xin, K.R. (2000) Guanxi in the Chinese context. In J.T.Li, A.S. Tsui and E. Weldon (eds) *Management and Organizations in the Chinese Context*. London: Macmillan Press.

Tushman, M.L. (1977) Special boundary roles in the innovation process. *Administrative Science Quarterly* 22, 587–605.

Tyler, D. and Dinan, C. (2001) The role of interest groups in England's emerging tourism policy network. *Current Issues in Tourism*, 4 (2–4), 210–52.

UNWTO. (2002) Tourism proves as a resilient and stable economic sector. Retrieved July, 2004, from http://www.world–tourism.org/newsroom/Releases/more_releases/june2002/data.htm.

Urry, J. (1990) *The Tourist Gaze: Leisure and Travel in Contemporary Societies*. London: Sage.

Urry, J. (1992) The tourist gaze 'revisited'. *The American Behavioral Scientist (1986–1994)* 36 (2), 172–86.

Uzzi, B. (1996) The sources and consequences of embeddedness for the economic performance of organisations: The network effect. *American Sociological Review* 61, 674–98.

Uzzi, B. (1997) Social structure and competition in interfirm networks: The paradox of embeddedness. *Administrative Science Quarterly* 42, 35–67.

Uzzi, B. (1998) Structural embeddedness and the persistence of repeated ties. Presented at the Academy of Management conference, August 9–13. San Diego, CA.

van Waarden, F. (1992) Dimensions and types of policy networks. *European Journal of Policy Research* 21, 29–52.

Verbole, A. (2000) Actors, discourses and interfaces of rural tourism development at the local community level in Slovenia: Social and political dimensions of the rural tourism development process. *Journal of Sustainable Tourism* 8 (6), 479–90.

Verbole, A. (2003) Networking and partnership building for rural tourism development. In D. Hall, L. Roberts and M. Mitchell (eds) *New Directions in Rural Tourism*. Burlington: Ashgate

Vernon, J., Essex, S., Pinder, D. and Curry, K. (2005) Collaborative policymaking: Local sustainable projects. *Annals of Tourism Research* 32 (2). 325–45.

Waldrop, M. (1992) *Complexity: The Emerging Science and the Edge of Order and Chaos*. London: Simon and Schuster.

Walker, J. (2002) Links and power: The political economy of linking on the Web. *Proceedings of the 2002 ACM Hypertext Conference, Baltimore, MD*, 72–73.

Walker, P.A., Greiner, R., McDonald, D. and Lyne, V. (1998) The Tourism Futures Simulator: A systems thinking approach. *Environmental Modelling and Software* 14 (1), 59–67.

Walle, A.H. (1997) Quantitative versus qualitative tourism research. *Annals of Tourism Research* 24 (3), 524–36.

Wang, Y. and Fesenmaier, D.R. (2007) Collaborative destination marketing: A case study of Elkhart county, Indiana. *Tourism Management* 28 (3), 863–75.

Ward, R.A. (1964) *Operational Research in Local Government*. London: Allen and Unwin.

Wasserman, S. and Faust, K. (1994) *Social Network Analysis*. Cambridge: Cambridge University Press.

Watts, D. (2004) The 'new' science of networks. *Annual Review of Sociology* 30, 243–70.

Watts, D.J. and Strogatz, S.H. (1998) Collective dynamics of 'small world' networks. *Nature* 393, 440–2.

Wearing, B. and Wearing, S. (1996) Refocussing the tourist experience: The flaneur and the choraster. *Leisure Studies* 15 (4), 229–43.

Weber, M (1947) *Social and Economic Organizations*. New York: Free Press.

Weber, M. (1947) *The Theory of Ssocial and Economic Organization*. London: Hodge.

Webster, C.M. and Morrison, P.D. (2004) Network analysis in marketing. *Australasian Marketing Journal* 12 (2), 8–18.

Welch, D.E., Welch, L.S., Young, L.C. and Wilkinson, I.F. (1998) The importance of networks in export promotion: Policy issues. *Journal of International Marketing* 6 (4), 66–82.

Wellman, B. (1988) Structural analysis: From method and metaphor to theory and substance. In B. Wellman and S.D. Berkowitz (eds) *Social Structures: A Network Approach*. Cambridge: Cambridge University Press.

Wellman, B. (2001) Computer networks as social networks. *Science* 293, 2031–4.

Werthner, H. and Klein, S. (1999) *Information Technology and Tourism: A Challenging Relationship*. Wien: Springer.

White, L. (2001) Effective governance through complexity thinking and management science. *Systems Research and Behavioral Science* 18, 241–57.

Wilenski, P. (1977) *Directions for Change: An Interim Report*. Review of NSW Government Administration. Sydney: Government Printer.

Wilenski, P. (1982) *Further Report: Unfinished Agenda*. Review of New South Wales Government Administration. Sydney: Government Printer.

Wilkinson, I.F. (2001) A history of network and channels thinking in marketing in the 20th Century. *Australasian Marketing Journal* 9 (2), 23–52.

Wilkinson, I. and Young, L. (2002) On cooperating firms, relations and networks. *Journal of Business Research* 55, 123–32.

Williams, J.D., Han, S.L. and Qaulls, W.J. (1998) A conceptual model and study of cross-cultural business relationships. *Journal of Business Research* 42, 135–43.

Williamson, O.E. and Ouchi, W.G. (1981) The markets and hierarchies progam of research: Origins, implications, prospects. In A.H. Van de Ven and W.F. Joyce (eds) *Perspectives on Organization Design and Behavior* (pp. 347–370). New York: John Wiley and Sons.

Wolfram, S. (2002) *A New Kind of Science*. Champaign, IL: Wolfram Media.

Wong, Y. and Tam, J. (2000) Mapping relationships in China: Guanxi dynamic approach. *Journal of Business and Industrial Marketing* 15, 57–70.

Wood, D.J., and Gray, B. (1991) Toward a comprehensive theory of collaboration. *The Journal of Applied Behavioral Science* 27 (2), 139–62.

World Tourism Organisation (2003) *Chinese Outbound Tourism*. Madrid: WTO.

Wright, M. (1988) Policy community, policy networks and comparative industrial policies. *Political Studies* XXXVI, 593–612.

Yang, M. (1986) *The Art of Social Relationships and Exchange in China*. Berkeley, CA: University of California.

Yang, M. (1994) *Gifts, Favors and Banquets: The Art of Social Relationships in China*. Ithaca, NY: Cornell University Press.

Yasin, M., Alavi J., Sobral F. and Lisboa J. (2003) Realities, threats and opportunities facing the Portuguese tourism industry. *International Journal of Contemporary Hospitality Management* 15 (4), 221–5.

Yau, O.H.M., Lee, J.S.Y., Chow, R.P.M., Sin, L.Y.M. and Tse, A.C.B. (2000) Relationship marketing the Chinese way. *Business Horizons* 43, 16–24.

Yin, R.K. (1994) *Case Study Research, Design and Methods* (2nd edn). London: Sage.

Yüksel, F., Bramwell, B. and Yüksel, A. (2005) Centralised and decentralised tourism governance in Turkey. *Annals of Tourism Research* 32 (4), 859–856.

Zeitz, G. (1980) Interorganizational dialectics. *Administrative Science Quarterly* 25 (1), 72–88.

Zipf, G.K. (1949) *Human Behavior and the Principle of Least Effort: An Introduction to Human Ecology*. Cambridge: Addison-Wesley.

Index